JULIAN HAWTHORNE

Incredible Mysteries: Ghost Lights

Contents

Introduction

Atmospheric ghost lights, a phenomenon as entrancing as it is enigmatic, have long held a place in human wonder and speculation. These elusive lights, often observed as they hover or flicker in isolated locations, have given rise to an array of myths and scientific hypotheses. In this exploration, we journey into the world of these spectral lights, examining their historical roots, scientific interpretations, cultural impacts, and the perpetual intrigue they generate among both devotees and doubters.

Historically, these ghost lights have been noted by diverse cultures across the globe. In European narratives, they were frequently regarded as supernatural signs or spirits. The English folklore's Will-o'-the-wisp, infamous for leading wayfarers astray, is a classic example. In contrast, Native American lore in the Americas often portrayed these lights as spiritual entities or indicators of hallowed grounds.

In the context of Asia, particularly in Japan, such lights are termed "hito-dama," symbolizing the spirits of those recently passed. Australian Aboriginal traditions, meanwhile, view them as celestial guides.

From a scientific standpoint, these ghost lights are frequently attributed to natural processes. A prevalent theory suggests the combustion of gases like methane or phosphine, typically emanating from decomposing organic material in wetlands. This phenomenon, referred to as "ignis fatuus" or "fool's fire," produces an illusion of dancing flames seemingly suspended in mid-air.

Additionally, geological factors are also considered. For instance, "earth lights" are hypothesized to stem from the geological strain along fault lines, generating electrical discharges that manifest as luminous spheres in the atmosphere.

Culturally, the impact of atmospheric ghost lights is significant and far-reaching. They have sparked a wealth of legends, myths, and folklore worldwide, serving not only as sources of entertainment but also as vehicles for imparting moral lessons. In the realms of literature and film, these ghost lights frequently appear as symbols of mystery, peril, or the supernatural.

In the present day, the allure of atmospheric ghost lights endures. They draw the attention of tourists, paranormal enthusiasts, and researchers alike. Advances in technology have facilitated more thorough investigations, yet numerous instances still defy explanation, adding to their enduring mystique.

I

North America

Baie Chaleur Fireship

In the captivating waters of Chaleur Bay in Canada, an enigmatic phenomenon known as The Chaleur Phantom unfolds, shrouded in mystery and intrigue. This ghostly light, emerging just before storms, has baffled observers and sparked a blend of scientific curiosity and folklore. Those equipped with telescopes, diligently studying this elusive apparition, find that even the most advanced optics cannot unravel its secrets. Up close, the Phantom remains just as cryptic, defying any concrete explanation.

To the naked eye, this spectral display takes on a more vivid form, resembling a ship engulfed in flames. This striking vision has fueled numerous tales and legends among the locals and visitors alike. Chaleur Bay, or "Baie des Chaleurs" in French, translates to "Bay of Warmth," a name that resonates with the fiery imagery of the ghost ship. This hauntingly beautiful bay, nestled between New Brunswick's rugged north shore and the scenic landscapes of Quebec's Gaspé Peninsula, has become the backdrop for this mysterious maritime ghost story. The Phantom ship, said to be ablaze in ethereal fire, is reported to roam these waters, leaving behind a trail of awe and speculation. This interplay of natural beauty, unexplained phenomena, and rich local lore makes Chaleur Bay a place of both warmth and haunting mystery.

The mysterious lights of The Chaleur Phantom in Chaleur Bay have been the subject of numerous tales and legends, each adding a layer of depth and intrigue to this enigmatic phenomenon. In the area west of Caraquet, the ghostly apparition is believed to be the Marquis de Malauze, a French vessel reportedly sunk by the British in the turbulent times of 1760. This story, steeped in the historical conflicts of the era, paints a vivid picture of maritime warfare and its haunting aftermath.

To the east, the Phantom takes on a different identity, known as the John Craig. This barque met its tragic end near Shippigan Island around 1800, leaving behind a tale of survival and sorrow. Among its crew, only a cabin boy narrowly escaped the fate of drowning, only to succumb later to exhaustion, adding a poignant chapter to the Phantom's lore.

Another captivating story attributes the spectral lights to the Lady Colbourne, a schooner that sank in 1838 along with its invaluable cargo. On its final voyage, the Lady Colbourne was laden with treasures like gold, silver, exotic spices, and fine wines, not all of which were recovered after the wreck. This ill-fated journey also claimed the lives of its affluent passengers, dressed in their finest attire, adding a layer of tragic opulence to the legend. The sinking of the Lady Colbourne is marked by the loss of 43 souls, making it a particularly somber tale associated with the Phantom.

Yet, these stories might not encapsulate the earliest origins of The Chaleur Phantom. A more ancient and perhaps the most heartrending tale is linked to Portuguese explorers. This narrative, woven into the fabric of local history, tells of their fateful end in the bay, marked by their contentious interactions with the indigenous people, including acts of enslavement.

On a tranquil summer evening in 1878, under the dimming sky of Heron Island, Mrs. Pettigrew found herself enveloped in an otherworldly encounter. Seated on her veranda as dusk settled, she was startled by the sudden appearance of a man, his form materializing before her, pleading for assistance. His features were marred by severe burns, evoking both sympathy and terror. In a moment of instinctive fear, Mrs. Pettigrew turned to flee inside her home. As the mysterious figure brushed past her, she was chilled by the realization that he had no legs. Yet, before she could process this eerie anomaly, he vanished into the ether, leaving behind a haunting mystery.

The Pettigrew family, long-time residents of Heron Island, had witnessed numerous unexplained phenomena over the years, particularly in the bay. They often spoke of a ghost ship, a spectral vessel that seemed to favor the north side of the island, appearing most frequently under the luminous glow of the full moon. This ghostly apparition had become a whispered legend among the islanders, a tale passed down through generations.

Among the numerous tales woven into the island's history, one particularly grim story stands out. It revolves around a Portuguese captain from the 1500s, infamous for his brutal raids and pillaging in the area, before mysteriously vanishing. This captain is believed to be Gaspar Cort-Real, a real historical figure who arrived at Heron Island in 1501. His mission was dark and mercenary: to kidnap the local Mi'kmaq people, native to the region, and sell them into slavery. Reports claim that he managed to capture as many as 57 indigenous individuals, who were then transported to Portugal to live out their lives in bondage.

However, on his second visit, the tables turned dramatically. The Mi'kmaq, bearing the painful memories of their previous encounter, captured Cort-Real. In a desperate attempt to prevent further atrocities, they tortured and eventually killed him, a brutal act of retribution for the suffering he had inflicted on their people.

The saga continued a year later with the arrival of Miguel, Cort-Real's brother, who came in search of him. He, too, faced the wrath of the locals. In an act of fierce defiance, they set his ship ablaze. According to legend, as they leapt into the water to escape the inferno, they vowed to haunt the bay for a millennium, giving rise to the legend of The Chaleur Phantom.

It is said that the shores of Heron Island bore witness to the grim aftermath of these encounters. The bodies of both the Portuguese invaders and the

Mi'kmaq warriors were found washed ashore. They were hastily buried in a low-lying area on the western tip of the island, known ominously as French Woods. The graves, shallow and hastily dug, were believed to be restless. The local lore suggests that the unrest of these souls, neither at peace in life nor in death, contributes to the eerie phenomena that continue to shroud Heron Island, including the mysterious apparitions and the legend of The Chaleur Phantom.

One particularly haunting tale emerges from the shores of Restigouche. This narrative, steeped in both tragedy and the supernatural, speaks of a time long past, where the ruthless deeds of pirates near Port Daniel set the stage for a curse that would echo through the ages.

The central figure of this tale is a woman, often described in the stories as a native of the land. She found herself in the clutches of a band of merciless pirates, who, driven by greed and devoid of empathy, kidnapped her. The circumstances of her abduction paint a vivid picture of the lawlessness and brutality that characterized the era, a time when the lives of the indigenous people were frequently caught in the crossfires of conflict and conquest.

As the story goes, the fate of this woman was as tragic as it was unjust. The pirates, having carried out their vile intentions, left her to meet a cruel end. In her final moments, consumed by both anguish and a desire for justice, she summoned the remnants of her strength to utter a curse upon her assailants. Her words, imbued with the weight of her suffering and the power of her spirit, were a chilling pronouncement:

"For as long as the world is, may you burn on the bay."

This curse, according to the legend, was no mere dying declaration. It was a potent spell that transcended time and death. The phantom lights that continue to flicker and dance across Chaleur Bay are said to be a manifestation of this curse. They are not just natural phenomena; they are believed to be the eternal flames of retribution, burning ceaselessly as a reminder of the woman's agony and the pirates' inhumanity.

Within the myriad of legends surrounding The Chaleur Phantom, a third narrative weaves its way into the lore, steeped in maritime superstition and tragedy. This tale speaks of a ship, its crew embattled by relentless and treacherous weather, which pushed them to the brink of despair. The sea's unpredictability and the crew's mounting desperation birthed a dangerous superstition among them. One sailor, in particular, was overwhelmed by a foreboding sense of doom, convinced that an ominous curse of bad luck was trailing their vessel, a belief that quickly infected the minds of his shipmates.

In this climate of fear and suspicion, the crew's paranoia focused on one of their own, a sailor whom they irrationally blamed for their misfortunes. Believing that his presence was the root of their ill luck, they made a fateful decision. In a misguided attempt to reverse their fortunes and appease the wrathful sea, they committed a heinous act - the murder of their fellow sailor.

However, the story takes a grim turn. Following the murder, the ship was engulfed in flames, a disaster that local lore attributes to the vengeful spirit of the wronged sailor. It was said that his Catholic blood cried out for retribution, and the fire that consumed the ship was the manifestation of this supernatural justice.

This tale is but one of many that have been spun in an effort to explain the enigmatic phenomenon of The Chaleur Phantom. The bay's mysterious lights have captivated both the scientific community and paranormal enthusiasts alike, leading to a multitude of theories and investigations. Researchers have dedicated numerous studies to demystifying these lights, offering explanations that range from the supernatural to the scientific.

Despite these efforts, factual inconsistencies in the ghostly tales and a scarcity of photographic evidence of The Chaleur Phantom have made it challenging to substantiate or debunk the legends. However, more natural explanations have been proposed, such as the emission of marsh gas from decomposing vegetation or an underwater release of natural gas. Another theory is the occurrence of St. Elmo's Fire, a weather phenomenon involving the discharge of atmospheric electricity. Although many scientists dispute its connection to The Chaleur Phantom, primarily because St. Elmo's Fire typically manifests as glowing tips on pointed objects and is often accompanied by a crackling sound.

Regardless of the true nature of The Chaleur Phantom's lights, their existence is an undeniable facet of the bay's mystique. Equally undeniable are the tragic and haunting stories that have emerged from the bay, each adding to the rich tapestry of ghost stories and local legends. These narratives, whether rooted in fact or fiction, continue to capture the imagination and speak to the timeless allure of the unknown and the unexplained.

The Ghost Ship of Northumberland

For centuries, the mystifying saga of a Ghost Ship has captivated the imaginations of those near the Northumberland Strait, an enigmatic waterway cradled by Nova Scotia, New Brunswick, and Prince Edward Island. This spectral vessel, steeped in folklore and mystery, is no ordinary apparition. Witnesses describe it as a stunning schooner, boasting three majestic masts. What sets it apart is the mesmerizing, yet chilling spectacle of the entire ship being consumed by roaring flames, a sight that leaves onlookers in awe and trepidation.

This maritime phantom isn't just a visual marvel; it carries with it ominous portents. The appearance of the ship is often linked to the arrival of a northeast wind, a harbinger of impending storms. This belief is deeply rooted in local folklore, suggesting that the blazing ghost ship is not merely a specter, but a warning sign from the depths of the ocean.

The chronicles of sightings span across the seasons, with a peculiar concentration from September to November. However, encounters are not confined to autumn alone. A particularly notable sighting occurred in mid-January 2008. Mathieu Giguere, a 17-year-old at the time, was captivated by the sight of a "bright white and gold ship" while gazing across Tatamagouche Bay. His account, which caught the attention of the Truro Daily News, added a modern chapter to this age-old legend.

The experiences of those who've encountered the Ghost Ship are not only visual but also auditory. The eerie prelude often includes the distant sound of cannon fire, followed by a sudden eruption of a massive flame, visible from all surrounding shores. As the spectral schooner materializes, witnesses report a frantic scene unfolding: the ship's crew, ghostly figures in their own right, desperately attempting to douse the inferno that engulfs their vessel. Yet, their efforts are in vain, as the flames seem impervious to their actions.

This legendary Ghost Ship of the Northumberland Strait, with its fiery masts and doomed crew, continues to be an enigmatic presence, a haunting blend of beauty and foreboding, forever sailing in the realms of maritime myth and local lore.

Over the years, there have been numerous tales of brave souls who, upon witnessing this phantom vessel ablaze, have embarked on daring rescue missions, driven by a sense of duty to aid those they believed were in peril.

One such notable episode is steeped in the maritime history of Charlottetown Harbour, dating back to around 1900. In this dramatic instance, a band of valiant sailors, spurred by the sight of the fiery ship, launched an impromptu rescue operation. They swiftly boarded a small rowboat, their hearts set on saving the crew they presumed to be trapped in the fiery inferno. With determination fueling their strokes, they raced towards the burning specter, battling the waves and their own mounting apprehension.

However, as they neared what they thought was a distressed vessel, the ghostly ship performed its most mystifying act: it vanished into thin air, leaving the would-be rescuers in a state of bewildered shock. Undeterred, and with a sense of duty still compelling them, the sailors initiated an exhaustive search. Divers plunged into the depths, scouring the seabed for any trace of a wreck, but their efforts were in vain. The waters revealed no secrets, no remnants of the phantom that had so convincingly appeared to be in distress.

This ghost ship, shrouded in mystery and legend, has sparked various theories about its origins. Among the more captivating is the belief that it was once a pirate ship, doomed to a watery grave by the relentless pursuit of the British Navy. In certain circles, especially among those who have witnessed its apparition in the Bay of Chaleur, Nova Scotia, the ship has been hauntingly dubbed "The Chaleur Phantom." This name resonates with the eerie, otherworldly nature of the ship and the chilling tales that have been woven around its ghostly appearances.

The earliest recorded sighting of this ghostly vessel dates back to the year 1786, a tale that has been passed down through generations, tinged with awe and an eerie sense of the supernatural.

This initial encounter occurred at the Sea Cow Head Lighthouse, an isolated beacon standing sentinel over a stretch of perilous waters. The lighthouse keeper, a man accustomed to the lonely vigilance of his post, was confronted with a sight that would forever be etched in his memory. In the midst of a raging northeast gale, a three-masted schooner emerged, its sails billowing fiercely in the storm's unrelenting grip. The ship, seemingly at the mercy of the tempest, was driven perilously close to the jagged rocks at the cliff's base, a sight that filled the keeper with a sense of impending doom. Yet, in a moment that defied all expectations, the vessel abruptly turned into the heart of the storm, disappearing into a veil of rain and mist, leaving the keeper in a state of shock and bewilderment.

Fast forward to January 1988, and the Ghost Ship made its presence known once again. This time, its burning silhouette was spotted off Borden, a startling vision against the night sky. Witnesses aboard a ferry observed the fiery apparition with a mix of fascination and fear. In an attempt to understand this enigma, the ship's radar was directed towards the mysterious sight. Yet, in a twist that deepened the mystery, the radar failed to detect any physical presence. The Ghost Ship remained an elusive phantom, visible to the eye but impervious to the tools of modern technology.

Captain Angus Brown of Wood Islands, a seasoned mariner, had his own chilling encounter with this maritime specter. Recalling an incident from his past, he spoke of a night when he and his crew aboard the ferry Prince Nova set out from Wood Islands with a mission to assist what appeared to be a burning ship. As they drew closer, ready to provide aid, the ship abruptly vanished into the night, leaving them in stunned silence.

On that same night, a couple residing near Glengarry witnessed the full-rigged ship ablaze from their bedroom window. Its sails illuminated by flames, the ship cut through the waters at an unfathomable speed, heading northward. The couple, familiar with the tales of the Ghost Ship, did not sound an alarm. Instead, they watched in silent awe, knowing they were witnessing a piece of living folklore. For over an hour, they observed the fiery vessel's journey across the sea, a private spectacle of the supernatural that they would remember for a lifetime.

Among the many accounts, one particular incident stands out, involving a group of men including the late John P. MacLean, a respected figure in the port of Charlottetown. This experience, transcending the ordinary, left an indelible mark on all who witnessed it.

On that fateful day, as the men were engaged in their routine tasks at the port, an urgent message swept through the area, stirring a sense of alarm and immediacy. A ship was reported in dire straits within the confines of Charlottetown Harbour. Eyewitnesses described a large three-masted sailing vessel, engulfed in flames from bow to stern. The scene was one of utter chaos and desperation as the ship's crew was seen scurrying frantically across the deck, engaged in a futile battle against the multitude of fires ravaging their vessel.

Compelled by a sense of duty and humanity, a rescue boat was quickly dispatched, its crew rowing with all their might towards the blazing ship, hoping against hope to save those aboard. However, the sea, as if conspiring with the fates, enshrouded the burning ship in a thick mist, obscuring it from view. By the time the rescuers arrived at the scene, the ship had disappeared, as if swallowed by the sea itself. Despite a thorough and relentless search, no trace of the ship or its crew was ever found, adding another layer of mystery to the already enigmatic tale.

The vivid recollection of an eyewitness provides a chilling and detailed account of the incident. As the ghostly ship approached, it seemed to slow down, coming to a complete halt opposite their house. The onlookers, filled with a mix of curiosity and trepidation, climbed onto the banks for a better view. Strangely, there was no sign of life on board, no crew visible on the deck, no lifeboats in tow.

The next moments were surreal and haunting. Smoke began to rise slowly from the deck, creating an eerie atmosphere. Suddenly, figures appeared, seemingly emerging from below deck, running in a state of panic. It wasn't long before a low flame enveloped the deck, consuming everything in its path. The desperate crew climbed the masts, trying to escape the inferno, but as they reached halfway, the sails caught fire. In an instant, the entire ship was ablaze, the flames obscuring the men from view.

The onlookers watched in horror as the fire raged, eventually dying down until all that remained was the ship's hull, eerily floating on the water. Then, in a gradual, almost spectral manner, the hull began to sink, as if being slowly claimed by the depths, until it vanished completely, leaving behind nothing but the chilling memory of its fiery demise.

One of the witnesses, still visibly shaken by the memory, recounted their experience: "It was a vessel, but unlike any I had seen before. Its outline was illuminated with a fiery glow, not like flames you could touch, but as if the whole ship radiated an otherworldly light. It moved across the water with such speed, it seemed almost unnatural." This ghostly apparition lingered for an astonishing two hours, its haunting image etched into the minds of dozens of onlookers along the road.

Rewinding to the autumn of 1973, another chapter in the tale of the Ghost Ship unfolds. Marvin MacLeod, a local resident, was driving home from Murray Harbour on a frosty, clear night. The chill in the air was sharp, and the sky was a canvas of stars. It was hunting season, and Marvin had a pair of binoculars in his car, a tool that would soon grant him an unexpected and startling view. Through the lenses, he saw what appeared to be a three-masted schooner, its sails seemingly engulfed in flames. Figures could be discerned on the deck, moving in chaos, some even leaping into the water in a desperate attempt to escape. Just as Marvin was about to race home to alert the Coast Guard, the ship disappeared right before his eyes, as if it had never been there at all.

In October 1978, an intriguing incident involving the last light keeper at Wood Islands, Leon Patton, added yet another layer to the legend. Leon received an unexpected phone call from Bernice Smith, a neighbor living up the road. Her query was simple yet puzzling: "What are you burning?" Confused, Leon replied that he was not burning anything. Urged by Bernice to look out his window facing east, Leon was met with a sight that defied explanation. There, moving down the Strait towards Murray Harbour, was the phantom ship, a schooner completely engulfed in flames. As he watched, the ship performed its most enigmatic trick; it vanished into thin air, leaving Leon and Bernice in a state of awe and disbelief.

Among those who have experienced this mysterious phenomenon is Larry Hooper of Murray River, who recalls distinct encounters with the spectral ship.

His first sighting dates back to the early 1980s, during an October night. While on lookout on the bridge of the Prince Nova around 10:00 p.m., near Pictou Island, he witnessed an astonishing sight. "It was a clear night, the kind where visibility stretches for miles. Suddenly, there it was – a ship with its sails engulfed in flames, and figures moving frantically on the deck. Then, as quickly as it appeared, it vanished," he recounts. His second experience occurred in the late 1990s, while driving with his uncle, Haldon, on Norman's Road. Once again, the burning ship materialized, with sailors visibly scrambling on its deck, before disappearing into the night.

Another local shared their encounter, "One October night, returning from a neighbor's, I gazed out over the Northumberland Strait. There, against the clear sky, was a burning ship. Its outline was distinct, and I watched in awe for about twenty minutes before it simply faded away. Having heard tales of the phantom ship, I knew then what I had witnessed."

A particularly vivid sighting from November 26, 1965, by a resident near Cape John, adds depth to these accounts. "It was early evening, just turning dark. I was near my kitchen window when I saw it – a ship on fire, sailing down the Strait. I immediately called my neighbors. Many from up the Cape and as far as River John, six miles away, saw it too. Crowds gathered, and the Phantom Ship was visible to hundreds for about half an hour before it faded into the darkness. Two nights later, under similar conditions, it appeared again, drawing even larger crowds to witness its eerie glow."

Roland Sherwood, an author who dedicated over 40 years to researching this phenomenon, compiled numerous witness accounts in his book, "The Phantom Ship." Sherwood himself had two encounters with the phantom, one at Caribou and another near Wallace, adding his own experiences to the growing body of evidence.

Despite some theories suggesting optical illusions or methane gas burning on the water as explanations, the sheer number of detailed and consistent accounts over the years lends credence to the possibility of a supernatural occurrence. These sightings, rich in detail and supported by dates and multiple eyewitnesses, continue to fuel the belief that the Ghost Ship of the Northumberland Strait is more than just a figment of collective imagination, but rather a genuine, unexplained marvel of maritime folklore.

Marfa Lights

The "Marfa Lights" of West Texas have been surrounded by an aura of mystery and intrigue for many years. These lights, known by various names such as ghost lights, weird lights, strange lights, and mystery lights, have been a subject of fascination and speculation. The most popular viewpoint for observing these enigmatic lights is a widened shoulder on Highway 90, located about nine miles east of Marfa. These lights, which are often seen as distant spots of brightness, are distinct from the ranch lights and automobile headlights on Highway 67 due to their unusual and erratic movements.

Further adding to the mystique of the "Marfa Lights," they are often described as being seen in a specific area to the south-southwest of the Marfa Lights Viewing Center (MLVC). The area for viewing is marked by distinct landmarks, with one margin aligned along a prominent telephone company tower and the other margin marked by the notable Chinati Peak, as observed from the MLVC.

In more detailed descriptions, the lights are said to appear as mysterious orbs that suddenly emerge above the desert foliage near Marfa Lights View Park. These orbs can exhibit a range of behaviors, from remaining stationary and pulsating to darting across the desert or performing complex movements like splits and mergers. The colors of these lights vary, encompassing hues of yellow-orange, green, blue, and red. Known as the Marfa Mystery Lights, they typically fly above desert vegetation but below the mesas, further contributing to their intriguing and mysterious nature.

The first documented sighting dates back to 1883, when a young cowhand named Robert Reed Ellison encountered a flickering light while herding cattle through Paisano Pass. Mistaking it for an Apache campfire, Ellison's curiosity was piqued, but further investigation by him and other settlers revealed no traces of a campsite, such as ashes, fueling the mystery.

These lights became a topic of local lore, with additional sightings reported by settlers like Joe and Anne Humphreys in 1885. These early accounts are captured in Cecilia Thompson's seminal work, "History of Marfa and Presidio County, Texas 1535–1946," published a century later in 1985. This book chronicles the tales and legends surrounding the lights, weaving them into the fabric of local history.

The intrigue surrounding the Marfa Lights gained national attention with the first published account in the July 1957 issue of Coronet magazine. This brought the phenomenon into the public eye, sparking curiosity and debate. Elton Miles further contributed to the lore in 1976 with his book "Tales of the Big Bend," which includes narratives dating back to the 19th century and even features a photograph of the Marfa lights taken by a local rancher, offering a tantalizing glimpse into this unexplained mystery.

In a more scientific approach to understanding these lights, extensive monitoring began in 2003. A significant compilation of sightings, numbering 34 from the period of 1945 through 2008, was documented by a dedicated observer. This meticulous record-keeping revealed an average of 9.5 sightings of the Marfa Lights on about 5.25 nights each year. However, it's believed that even this diligent monitoring might only be capturing half of the occurrences in the area known as Mitchell Flat, suggesting that the phenomenon might be more frequent than recorded.

The Marfa Lights have been the subject of various interpretations and investigations, ranging from the scientific to the speculative. Brian Dunning, a notable skeptic, offers an intriguing perspective. He points out that the designated viewing area for these lights, a roadside park on U.S. Route 90 approximately 9 miles east of Marfa, coincides with the location of the former Marfa Army Airfield. This airfield, operational between 1942 and 1947, was a bustling hub where tens of thousands of personnel were stationed, training American and Allied pilots. Later, it served as a regional airport with daily airline service. Given the constant patrol by sentries in these military and aviation fields, Dunning posits that it's improbable for any unusual phenomena to have gone unnoticed. He suggests that the Marfa Lights could be a type of mirage, resulting from sharp temperature gradients in the air, a plausible explanation given Marfa's high elevation and significant temperature differentials.

Further adding to the debate, a group from the Society of Physics Students at the University of Texas at Dallas undertook a meticulous investigation in May 2004. Over four days, they used traffic volume-monitoring equipment, video cameras, binoculars, and chase cars to observe the lights. Their findings are intriguing: the frequency of lights observed correlated with vehicle traffic on U.S. Highway 67, and the motion of these lights was consistent with the direction of the highway. Experiments with parked cars and flashing headlights reinforced their conclusion that the observed lights could be reliably attributed to automobile headlights traveling along the route between Marfa and Presidio, Texas.

In a scientific endeavor to unravel this mystery, researchers from Texas State University conducted a spectroscopic analysis over 20 nights in May 2008. Observing from the Marfa lights viewing station, they recorded numerous lights that could have been mistaken for mysterious origins. However, the movements of these lights, alongside data from their equipment, pointed towards more mundane explanations like automobile headlights or small fires.

The Marfa Lights have also permeated popular culture, capturing the imagination of artists, writers, and filmmakers. They have been featured in television shows like "Unsolved Mysteries," "King of the Hill," and the Disney Channel's "So Weird." The phenomenon inspired David Morrell's 2009 book "The Shimmer," and even found its way into music, with the Rolling Stones referencing the "lights of Marfa" in their song "No Spare Parts." The country music artist Paul Cauthen wrote "Marfa Lights," a love song inspired by the phenomenon, for his 2016 album "My Gospel." Even the iconic Simpsons family encountered the Marfa Lights in a 2019 episode, adding to the lore and legend of these mysterious lights.

These varied interpretations and representations of the Marfa Lights highlight the enduring intrigue and fascination surrounding this phenomenon. From scientific analyses to cultural references, the lights continue to be a source of wonder, debate, and inspiration.

Gurdon Light

On the brisk morning of December 3rd, 1931, a startling incident shook the quaint town of Gurdon, Arkansas. A massive freight train, consisting of 14 heavy cars, tragically derailed just a few miles southwest of the town. This was no ordinary mishap. Upon a meticulous investigation, railroad officials unearthed a chilling fact: the rails had been deliberately tampered with. Their suspicions were dire – it appeared that the saboteur had intended a far more catastrophic target: the Sunshine Special. This renowned passenger train, known for its swift journeys at speeds between 60 and 70 mph, would have met a disastrous fate had it been the victim of this sinister plot.

The intrigue deepened the very next day, December 4th, 1931. The town was abuzz with unease when Will McClain's wife, fraught with worry, approached the city marshal. Her husband, a respected section foreman for the Missouri Pacific Railroad, had uncharacteristically failed to return home from work the previous night. Her anxiety was palpable; she had scoured the small community since dawn, only to find that no one had seen Will since his last shift at the railway. This was alarmingly out of character for the devoted foreman.

Seized by a sense of urgency, the marshal embarked on a thorough search. He canvassed the town, probing for any trace of Will, his inquiries painting a picture of growing concern among the locals. It was during this search that he encountered Louie McBride. McBride, a laborer known to work under McClain, struck the marshal as peculiarly evasive. Sensing something amiss, he acted decisively, detaining McBride for questioning. The town held its breath as the jail doors closed, wondering what secrets this interrogation might unveil in the shadow of the eerie train derailment.

In the early hours of a cold December morning in 1931, an alarming event unfolded near the quiet town of Gurdon, Arkansas. A freight train, burdened with 14 cars, suffered a catastrophic derailment just southwest of the town's outskirts. This was no ordinary accident. Railroad officials, upon inspecting the twisted wreckage, uncovered a disturbing truth: the rails had been maliciously tampered with. The implications were grave. It appeared that the target was the Sunshine Special, a beloved passenger train renowned for its rapid transit, often clocking speeds between 60 and 70 mph. Whoever was behind this act had seemingly intended for a tragedy of far greater magnitude.

The plot thickened the very next day, December 4th, 1931. The serene dawn was pierced by a wife's distress. Will McClain's spouse, fraught with worry, hastened to inform the city marshal of her husband's uncharacteristic disappearance. Will, a dedicated section foreman for the Missouri Pacific Railroad, had failed to return from his shift the previous night. Her concern was deepened by her fruitless inquiries within the community; nobody had laid eyes on Will since he was last seen at his workplace. It was utterly unlike him to not return home.

The city marshal, sensing the gravity of the situation, embarked on an exhaustive search throughout Gurdon. His mission was to unravel the mystery of Will's whereabouts. During his investigation, the marshal's attention was drawn to Louie McBride, a laborer known to be under Will McClain's supervision. McBride's demeanor was oddly suspicious, prompting the marshal to take him into custody for questioning.

Under the stern gaze of the law, McBride confessed to a heinous act. He admitted to brutally beating Will McClain to death, driven by a seething rage over being denied seniority rights amidst recent job layoffs. This revelation cast a dark shadow over the town, further complicated by the initial suspicion that McBride might have also been connected to the previous day's rail sabotage.

The trial was swift and decisive. Louie McBride was found guilty of murder and sentenced to the grim fate of death by electrocution. Despite numerous appeals, his fate was sealed. On July 8th, 1932, McBride faced his final moments in the electric chair at the Little Rock penitentiary, a stark ending to a tragic sequence of events that had begun with a train derailment and ended with the loss of two lives, shaking the foundations of the small Arkansas community.

Reports started flooding in, describing a peculiar, almost supernatural light along a four-mile expanse of the railroad track near the town. This mysterious illumination, varying in hues from a strange yellowish-white to an ethereal bluish tint, captivated the imagination of all who heard of it. The Gurdon Light, as it came to be known, hovered mysteriously a few feet above the railroad tracks, its shape reminiscent of a floating medicine ball.

This spectral light seemed to possess an almost playful, yet unnerving, character. Witnesses were both intrigued and unsettled as it vanished without a trace, only to reemerge suddenly, often materializing eerily behind those who walked the tracks in search of this ghostly apparition. Adding to the unsettling aura, the light's favored haunt was an area shrouded by dense, swampy woods, which whispered secrets of the unknown. Nearby, an old, small cemetery stood as a silent sentinel, its presence intensifying the eerie atmosphere.

As word of this mysterious light spread, it captured the curiosity of the town's residents. People began organizing excursions, venturing into the enigmatic area with a blend of excitement and trepidation, all hoping to catch a glimpse of this inexplicable phenomenon. By the 1950s, the Gurdon Light had transcended local folklore and captured the attention of the media. Articles in local newspapers began to chronicle the phenomenon, further fueling public interest.

Over the decades, the legend of the Gurdon Light grew. Countless individuals, from local enthusiasts to intrigued visitors, flocked to this mysterious stretch of railroad track. Each came with the hope of experiencing the unexplained, to witness firsthand the light that had become the subject of so many stories.

Over the years, many have endeavored to demystify this enigmatic phenomenon. Skeptics and scientists alike proposed various theories: some attributed the light to the reflections of headlights from a nearby highway; others speculated it to be "swamp gas" emanating from the earth. However, these explanations struggled to hold up against scrutiny.

One significant challenge to the headlight theory is historical. Reports of the Gurdon Light date back to the early 1930s, a time long before the construction of the nearby highway. Additionally, the geography of the area itself casts doubt on this theory. A substantial hill lies between the railroad tracks, where the light is most often seen, and the nearest highway, acting as a natural barrier that would obscure any such reflections.

Enter Mike Clingan, an individual whose dedication to unraveling the mystery of the Gurdon Light surpasses most. Clingan, with a background in physics, offers a compelling scientific hypothesis: the piezoelectric effect. This phenomenon occurs when stress is applied to certain materials, like quartz crystals, which are abundant beneath the tracks in Gurdon. This stress can generate an electrical charge, manifesting as the enigmatic glowing light observed by so many.

Gurdon's location at the edge of the New Madrid seismic zone lends credence to this theory, suggesting a geological explanation for the mysterious sightings. However, Clingan notes that the lore of the Gurdon Light predates even the tragic murder of Will McClain, a well-documented event that adds a chilling layer to the legend. Local "old-timers" recall stories of the light that date back to the era of the Native Americans and the earliest white settlers.

The Gurdon Light, unlike many ghostly phenomena, is steeped in rich narrative elements that make for a compelling ghost story. It combines a violent historical event, swampy, foreboding terrain, a nearby cemetery, and even reports of Bigfoot sightings along Stickey Road, adding layers of intrigue and suspense.

Clingan's personal experiences add to the enigma. As a physicist, he recalls observing unexplainable occurrences, such as watching groups of people approach along the tracks through binoculars, only for them to inexplicably vanish upon arrival. He describes witnessing a strange bluish mist hovering over the rails, along with disembodied voices and shuffling sounds with no discernible source. On one occasion, a profound sense of dread overcame him while rejoining his group, a feeling that was inexplicably shared by the others, despite no prior communication.

The Gurdon Light, therefore, stands at the crossroads of science and folklore. While a scientific explanation like the piezoelectric effect might account for its physical manifestation, the light also serves as a spectral emblem of the area's dark history, a reminder of a brutal murder that happened over 80 years ago. This confluence of science and story makes the Gurdon Light not just a natural phenomenon, but a legend woven into the very fabric of the local community.

Cohoke Light

Nestled in the historic landscapes of King William County, lying between the quaint towns of West Point and Danville, there's a place steeped in eerie lore - the old Southern Railroad line. Just off the beaten path of Mt. Olive Cohoke Road, you'll find the Cohoke Crossroads, a seemingly ordinary railroad crossing. However, it's shrouded in spine-tingling tales and ghostly whispers.

On nights when the fog rolls in thick, creating a blanket of mystery over the landscape, a bizarre and unsettling phenomenon occurs. A mysterious, glowing yellow light emerges on the train tracks. This spectral light, appearing on misty, gloomy nights, has a haunting quality to it. Sometimes it dances and flickers in the darkness, resembling a lantern swaying in the hands of an unseen, otherworldly figure. Other times, it remains ominously still, growing brighter and more intense as if a massive train engine is barreling down the tracks, ready to emerge from the fog.

The origin of this enigmatic Cohoke Light is shrouded in local folklore and chilling tales. Two predominant stories attempt to explain this supernatural occurrence. The first tale transports us back to the 1800s, a time when trains were the lifelines of America, and a variety of employees, each with their unique roles, maintained these steel beasts. Among them was the brakeman, a crucial figure responsible for ensuring the train's safety and functionality. He would patrol the train, lantern in hand, vigilantly checking for any anomalies, ready to signal the engineers in case of any hazards.

One fateful night, a brakeman performing his duties at the Cohoke Crossroad was inspecting his halted train. He discovered an issue with the coupling mechanism between two cars. Absorbed in his task, he failed to notice the train's sudden movement. In a horrific twist of fate, the train jolted forward, leading to a gruesome accident that claimed his life, decapitating him in the process. His colleagues, shocked and distraught, rushed to his aid, only to find his life had been tragically snuffed out. In the chaos and confusion, his head was never recovered.

This is where the legend takes a ghostly turn. It's said that the Cohoke Light is none other than the restless spirit of the headless brakeman, eternally roaming the tracks with his lantern. He's believed to be in an endless search for his lost head, a tragic soul condemned to wander the railway in the dead of night. This haunting tale has become a part of the local folklore, sending shivers down the spines of those who dare to venture near the Cohoke Crossroads after dark.

Another captivating theory weaves its way through the local folklore, suggesting a more dramatic and historical origin. This narrative takes us back to the tumultuous times of the Civil War, a period rife with conflict and desperation.

According to whispered tales and local historians, during the height of the Civil War, a critical incident occurred on these very tracks. A steam train, burdened with the solemn task of transporting wounded Confederate soldiers, embarked on a journey towards West Point. The goal was to evacuate these soldiers from the besieged city of Richmond, offering them a chance for survival and recovery. However, fate had a different plan. This train, filled with the hopes and pains of its passengers, mysteriously vanished before reaching its destination.

Local lore suggests a grim possibility: the train may have fallen victim to a Union ambush. In this tragic scenario, the Confederate soldiers aboard met a violent end, and the train itself was obliterated. The Cohoke Light, as per this legend, could be the spectral manifestation of that ill-fated train. It's as if the ghost of the steam engine, still ablaze with determination, continues its journey night after night, eternally trying to fulfill its mission to reach West Point, carrying the souls of its long-lost passengers.

Adding another layer to the mystery, some contemporary accounts propose a more otherworldly explanation. Could this strange, shape-shifting light be of extraterrestrial origin? Eyewitnesses have reported eerie variations in the light's appearance, noting changes in color and form. Typically presenting as a yellow glow, it occasionally morphs, displaying an uncanny spectrum of colors.

One particularly chilling account comes from a Richmond man and his four friends. On a dark, clear night at the crossroads, they encountered what they initially mistook for the glow of an arc welder. As they watched in awe, the light suddenly vanished, only to reappear with a startling change. It shifted from a singular red beacon to an array of lights, seemingly performing a spectral dance before vanishing into the night. This account has fueled speculation about the light's origin, with some firmly believing in its extraterrestrial roots.

The mystery often leads to the suggestion of a natural phenomenon as an explanation - specifically, the concept of swamp gas. This idea, though it may sound like the stuff of science fiction, is actually grounded in real scientific phenomena and has been a part of global folklore for centuries.

The legends surrounding swamp lights are diverse and rich, often featuring in tales where travelers in swamps or bogs are enticed by a mysterious light in the distance. Known by various names like will-o'-the-wisps or foxfires, these lights have a notorious reputation. In folklore, they're often depicted as malevolent, leading the unwary into perilous situations where they either vanish without a trace or meet a sinister end. While the mythical beings - demons, fairies, or ghosts - that are said to create these lights might be figments of the imagination, the lights themselves are a tangible reality in swampy terrains.

To understand this phenomenon, one must delve into the unique environmental conditions of bogs, swamps, and marshes. These areas are characterized by stagnant, non-flowing water, which becomes a breeding ground for decaying plant and animal matter. This decomposition process releases various gases, including methane, which are trapped under the water. In bogs, for instance, one might observe bubbles rising to the surface - a sign of methane being released into the air.

Methane, along with other gases produced in these environments, is highly flammable. Under certain conditions, when these gases come into contact with oxygen in the air, they can ignite, creating startling and unexpected bursts of flame. This phenomenon can manifest as sudden, brief fireballs, seemingly appearing out of nowhere.

This natural occurrence was even noted historically, such as in the diaries of soldiers during the Revolutionary War. Reports from these times mention instances of men hiding in swamps or bogs, only to have their clothing mysteriously catch fire, likely due to the spontaneous ignition of swamp gas.

Given this scientific background, it's plausible to consider swamp gas as a potential explanation for the Cohoke Light. However, like many other theories attempting to unravel the mystery, it struggles to fully account for all aspects of the phenomenon. The light's behavior, its consistent appearances, and the local context present challenges that swamp gas alone can't seem to explain.

Among those who have delved into this mystery is Bill Palmer, a local historian with a penchant for unraveling the secrets of local legends. Author of several books on regional folklore, Palmer has a particular interest in the tales surrounding the Cohoke Light, which he notes predominantly began to surface in the 1950s. This timeline, he argues, is curiously long for a ghostly haunting to commence, suggesting a more recent origin for the phenomenon.

Palmer scrutinizes the popular legend of the headless brakeman, a tale that, while compelling, lacks historical substantiation. Despite extensive searches, there are no death records from the 1800s that validate this grisly story. Furthermore, similar legends of tragic railroad workers appear in various other parts of the country, diluting the uniqueness of the Cohoke Crossing narrative. While tragic railroad accidents did occur, including a car accident on these very tracks involving a mother and daughter, none match the sensational nature of the decapitation story.

Turning to the theory of the missing Confederate train, Palmer is equally skeptical. He points out the lack of historical records supporting this event. Moreover, the narrative logistics don't align with the known movements of Union and Confederate forces during the Civil War. The story's suggestion that a train carrying wounded soldiers was heading south, directly into Union territory, contradicts established military strategies and historical accounts.

Palmer also addresses the swamp gas hypothesis. While it's a scientifically plausible explanation for mysterious lights, it fails to accurately describe the characteristics of the Cohoke Light. Swamp gas, typically resulting in erratic and fleeting bursts of flame, doesn't resemble a spherical, gently floating orb. Moreover, the Cohoke Light is known for its seemingly scheduled appearances, unlike the unpredictable and sporadic nature of swamp gas emissions.

The Cohoke Crossroads lore is filled with eerie encounters and personal accounts. Witnesses often report seeing a vague figure near the light that disappears quickly, and some even claim the light stops moving when confronted with gunfire. Despite attempts, capturing the Cohoke Light on camera remains elusive, leaving only stories and eyewitness accounts. These narratives continue to attract the curious and brave, especially on moonless nights. Visitors to these mysterious crossroads hope to glimpse the elusive light and perhaps add their own experiences to this enduring legend.

Maco Light

The Maco Light, a mysterious and captivating phenomenon, was occasionally observed from the late 19th century until 1977 along a stretch of railroad track near Maco Station, an unincorporated community nestled in Brunswick County, North Carolina. This enigmatic light, often described as resembling the warm, flickering glow of a railroad lantern, became the centerpiece of local folklore and a source of intrigue and speculation.

According to popular legend, the light was thought to be linked to a tragic and fatal accident on the railroad, a story that not only captivated the imaginations of locals but also may have inspired similar ghostly tales across the United States. This spectral light, dancing along the tracks in the still of the night, became a symbol of mystery and a testament to the power of local lore in shaping the cultural narrative of a region.

Despite its storied past, the true nature of the Maco Light remained a topic of debate and speculation. Some theorized that it was a natural phenomenon, perhaps the result of marsh gas emissions from nearby swamps, casting an eerie, otherworldly glow. Others suggested it could be a trick of light, a mere refraction from distant highway lights, playing on the observer's eyes and imagination. Yet, despite these theories, the Maco Light retained its mystique, a ghostly beacon that continued to allure and mystify those who heard its tale or dared to witness its fleeting, mysterious presence.

The legend of the Maco Light intertwines with the tragic tale of Joe Baldwin, a train conductor whose life met a grisly end in a harrowing railroad accident near Maco, North Carolina. This story, deeply embedded in local folklore, centers around an event that supposedly took place in the late 1800s along the Wilmington and Manchester Railroad.

The most widely recounted version of this ghostly tale begins on a stormy night in 1867. Joe Baldwin was aboard a train bound for Wilmington when, in a stroke of misfortune, the rear car he was in became detached from the main train. Alone and aware of the imminent danger, Baldwin realized another train was following closely behind. In a desperate attempt to avert disaster, he rushed to the rear platform of the dislodged car, lantern in hand, frantically waving it to signal the approaching train. Tragically, his efforts were in vain. The oncoming locomotive failed to notice the stranded car in time, resulting in a catastrophic collision that decapitated Baldwin. Adding a macabre twist to the story, some versions suggest that Baldwin's head was never recovered, leaving an unsettling mystery in its wake.

In the aftermath of this tragic incident, residents of Maco and railroad employees began to report sightings of a mysterious white light along a specific stretch of the railroad track, weaving through the swamps west of Maco Station. These sightings soon gave rise to the belief that the spirit of Joe Baldwin had returned, eternally searching for his lost head. Witnesses described the light appearing in the distance, then eerily advancing along the tracks, bobbing about five feet off the ground. Some accounts depicted the light veering off to the side of the track in an arc, while others said it would retreat from the viewer. There were also reports of the light displaying different colors, including green and red, and exhibiting varied patterns of movement.

The earliest recorded accounts of these strange phenomena date back to the 1870s, and the sightings became even more prominent following the 1886 Charleston earthquake. It was during this period that two lights were frequently reported, leading to interruptions and delays in train services, as the mysterious light occasionally appeared right in front of locomotive cabs.

The tale of the Maco Light gained such prominence that it even reached the ears of President Cleveland during a train stop at Maco in 1889, as recounted by Atlantic Coast Line employee B. M. Jones, who claimed to have been a young witness at the time. Further adding to the legend's notoriety, the journal Railroad Telegrapher reported a sighting as recent as March 3, 1946, suggesting the light had been a recurring phenomenon for nearly seventy years.

This captivating story of the Maco Light and Joe Baldwin has not only become a cornerstone of local folklore but also seems to have influenced similar "ghost light" legends across the United States. Tales like the Bragg Road ghost light and Gurdon light bear striking resemblances, suggesting that the Maco Light's story, with its national coverage and enduring intrigue, may have served as the inspiration for these other ghostly phenomena. As such, the legend of Joe Baldwin and the Maco Light stands as a fascinating example of how local folklore can capture the imagination and resonate through generations, weaving a tapestry of mystery and intrigue into the cultural fabric of a region.

The legend of the Maco Light evolved into a captivating regional sensation, drawing scores of intrigued individuals and groups, each seeking to unravel the mystery of this enigmatic phenomenon. Among these were a team of electronics engineers — two from radio station WWOK, one from WKIX, and another from the renowned Bell Laboratories — who ventured to the site in July 1962, armed with the latest technology and a drive to demystify the elusive light.

During the 1950s and 60s, the tracks near Maco Station transformed into a popular local haunt, especially at night. People would gather, parking their cars close to the railroad, their eyes fixed on the horizon, hoping to catch a glimpse of the ghostly illumination. Such was the allure of the Maco Light that it even caught the attention of the national media. Life magazine, captivated by the tale, dedicated a two-page article to the phenomenon in its October 1957 issue, further fueling public interest.

Photographers from the Wilmington Star-News embarked on their own quests to capture this spectral light, with attempts in 1946 and 1955. Although they claimed only partial success, their efforts underscored the challenging and elusive nature of the light. Even renowned personalities were drawn to the mystery; Joseph Dunninger, a famous magician and skeptic, visited Maco in 1957 but left without encountering the phenomenon. In 1964, paranormal researcher Hans Holzer conducted an investigation, leading him to a poignant theory — despite not witnessing the light himself, he speculated that the spirit of Baldwin was unaware of his death and continued his vigilant efforts to warn oncoming trains of danger.

Amidst these explorations and theories, locals offered more grounded explanations. Some attributed the light to phosphorescent gas emanating from the swamp or the reflection of distant car headlights. Others noted that the light was a regular occurrence until 1935, when the railroad undertook a project to fill in the swamp under a nearby trestle. Following this alteration, sightings of the light supposedly diminished, with only automobile headlights visible thereafter.

The mystery deepened with a revelation from the Wilmington Railroad Museum. A thorough search of newspaper archives revealed no record of an accident in 1867 involving a Joe Baldwin. However, they did uncover details of an incident in January 1856, involving a conductor named Charles Baldwin. This accident, which occurred close to what would later become Maco Station, happened when a locomotive, after detaching from its train to resolve technical issues, collided with it upon return. Charles Baldwin, though not decapitated, was fatally injured and later succumbed to his wounds. Interestingly, the coroner's report blamed Baldwin for not placing a lamp on the train to signal the engineer.

This discovery suggested a possible origin for the Joe Baldwin legend. Charles Baldwin, known to have been well-regarded in the community as evidenced by his obituary in the Wilmington Journal, might have inadvertently become the central figure in the Maco Light legend through a distortion of memories over time.

In 1972, the Wilmington Star-News published an article that added a new dimension to the ongoing debate about the Maco Light. This piece posited a theory that resonated with many investigators: the mysterious light was likely an optical illusion caused by the refraction of car headlights from the nearby U.S. Route 74. The article referenced a long-exposure photograph from 1950, which seemed to support this hypothesis. It showed how a bend in the highway could create an optical illusion, producing the enigmatic light. Notably, the article pointed out that amber and red lights, akin to truck turn and brake signals, were sometimes observed close to the main light when viewed through a telescope, adding weight to the car headlights theory.

The newspaper also noted a significant change in the frequency of sightings following highway modifications in the late 1960s, which included straightening the previously mentioned bend. This alteration, the article suggested, could explain the decreased occurrences of the light. In an attempt to address the claims of some locals, who insisted that the light continued to appear even when the highway was closed during World War II, the Star-News researcher conducted a meticulous review of archives. However, this investigation failed to uncover any evidence of such a closure, casting doubt on these assertions.

The Maco Light continued to be a topic of lively discussion in the community, often centered around the pages of the Wilmington Star-News. Some residents, aligning with the newspaper's findings, dismissed the supernatural aspects of the legend. One letter to the editor described the Maco Light as merely a lovers' lane, a place for mischief rather than a site of paranormal activity.

Despite the skepticism, the legend of the Maco Light maintained a cohort of believers and witnesses. In the mid-1970s, a reporter from The Robesonian, after several attempts, claimed to have finally seen the light. His description painted a chilling picture: a light akin to that from a kerosene lantern, predominantly white with a faint reddish tint, swinging and undulating slightly as it traveled down the center of the track.

Adding to the lore, author Bland Simpson, in a 2005 interview on North Carolina Public Radio, recounted his personal experience with the Maco Light. He described it as "one of [his] favorite" North Carolina legends, likening the light he saw to the flame of a match or the glow from a kerosene lantern. Simpson's account, like many others, was tinged with wonder and mystery, leaving the true nature of the light an open question.

The sightings of the Maco Light came to an end in 1977, coinciding with the removal of the railroad track and the destruction of a trestle bridge linked to the legend. However, the story of the Maco Light lives on, not only in the tales and memories of those who claim to have seen it but also in the local landscape. A street in a nearby subdivision, named Joe Baldwin Drive, serves as a lasting reminder of the enduring mystery and allure of the Maco Light, a legend deeply rooted in the folklore and history of North Carolina.

Paulding Light

The Paulding Light, also known as the Lights of Paulding or the Dog Meadow Light, is a mysterious and captivating phenomenon occurring in a serene valley located outside Paulding, Michigan. This enigmatic light has sparked curiosity and wonder since its first reported sightings in the 1960s. The area, enveloped in a rich tapestry of local folklore, has seen various explanations for this phenomenon ranging from the spectral – such as ghosts wandering the valley – to natural causes like unusual geologic activity or the emissions of swamp gas.

The precise location of this phenomenon is as picturesque as it is mysterious. The Paulding Light reveals itself in a tranquil valley near Watersmeet, located off US 45 on Robbins Pond Road/Old US 45, in the lush and forested landscape of Michigan's Upper Peninsula. This setting, away from the hustle and bustle of city life, offers a perfect backdrop for this enduring mystery, making it a destination for both thrill-seekers and those fascinated by the intersection of folklore and science.

The first recorded sighting of the enigmatic Paulding Light dates back to 1966, a year that marked the beginning of a captivating and mysterious narrative. It was a group of teenagers, with eyes wide in astonishment, who first brought this peculiar light to the attention of a local sheriff in Paulding, Michigan. Since this initial encounter, the light has become a staple of local lore, reportedly manifesting almost every night at the same site, drawing the curious and the brave to witness its spectral dance.

The stories woven around the Paulding Light are as varied as they are fascinating. Among these, the most prominent and haunting legend is that of a railroad brakeman's tragic demise. According to this chilling tale, the valley once echoed with the clatter of railroad tracks. It was here that a brakeman, in a valiant but doomed effort, met his end while trying to prevent a catastrophic collision between an oncoming train and railway cars stranded on the tracks. The light, they say, is the ghostly lantern he carried, continuing his eternal vigil in the afterlife.

Another spine-tingling story tells of a slain mail courier, his spirit said to roam the valley, with the light being a manifestation of his ghostly presence. A more mystical tale speaks of a Native American spirit, dancing eternally upon the power lines that traverse the valley, with the light being an ethereal reflection of this spectral dance.

One particularly poignant legend, as recounted by John Carlisle of the Detroit Free Press, suggests the light is the manifestation of a grandparent's unending search for a lost grandchild. This tale, tinged with sorrow and hope, proposes that the light's intermittent appearance is due to the grandparent relighting their lantern, a symbol of their unwavering hope and enduring love, as they continue their ceaseless search through the ages.

Each of these tales, whether grounded in tragedy, mystery, or enduring love, contributes to the rich tapestry of folklore surrounding the Paulding Light. They weave a narrative that not only captivates the imagination but also invites both skeptics and believers to the site, each seeking to unravel the truth behind this enduring Michigan mystery.

In October 1990, a team of diligent investigators, equipped with the tools of modern science such as telescopes, spectroscopic analysis, and meticulous travel time studies, embarked on a quest to demystify the Paulding Light. Their findings pointed to a rather mundane but fascinating explanation: the light was identified as the head and tail lights of vehicles traversing the north–south stretch of US 45, located approximately five miles north of where the light is observed. This revelation, although less spectral, opened the doors to understanding the intricate interplay of light and atmosphere.

Fast forward to 2010, when a group of astute students from the Michigan Tech chapter of the Society of Photo-Optical Instrumentation Engineers took up the mantle to further investigate this phenomenon. Through the eye of a telescope, they pierced the veil of night to observe not just the mysterious light, but also the mundane yet pivotal details of the landscape – vehicles moving, stationary objects, and even a specific Adopt a Highway sign. These students not only observed but also recreated various aspects of the light, such as the multicolored patterns resembling police flashers and the variations in intensity akin to high and low beams of a car. Their hypothesis? The unique stability of an atmospheric inversion layer, which acted like a lens, projecting the lights from a stretch of highway 4.5 miles away into the observing area.

Paranormal researcher Ben Radford offers a broader perspective on such mysterious lights, noting that they are not unique to Paulding but are reported across the United States. He acknowledges the plethora of potential sources for these lights – ranging from the mundane like headlights and campfires to the more unusual like aircraft, cloud reflections, or even insects. Radford highlights the human penchant for mystery and the allure of the unknown: while scientific explanations might reveal the origins of these lights, the romanticism of attributing them to spectral sources, like a ghostly railroad brakeman's lantern, often holds a more enchanting appeal than the headlights of a modern vehicle.

In this way, the Paulding Light stands at the crossroads of science and folklore, encapsulating the human fascination with the unknown and the relentless pursuit of truth. It is a testament to our desire to explore and understand the mysteries of the world around us, whether through the lens of a telescope or the lens of imagination.

Silver Cliff Cemetery

Silver Cliff Cemetery, nestled just a mile south of State Highway 96 on Mill Street in Colorado, presents a fascinating blend of history and mystery. Established in the early 1880s outside the quaint town of Silver Cliff, this cemetery carries a unique aura, steeped in the rich heritage of the area.

The cemetery is intriguingly divided into two distinct sections, reflecting the religious diversity of the community it serves. One half is dedicated to Catholic burials, resonating with traditional religious practices, while the other half, known as the Cross of the Assumption Cemetery, is reserved for Protestant burials. This division not only serves practical purposes but also mirrors the cultural tapestry of the town. Owned and operated by the Town of Silver Cliff, the cemetery continues to be an active burial ground, connecting the past with the present.

But what really sets Silver Cliff Cemetery apart are the mysterious "dancing blue lights" that have captivated the imagination of locals and visitors alike. These ethereal lights, often described as resembling blue lanterns or white spheres, have a storied history of floating through the graveyard and playfully bouncing off headstones. These spectral displays have been a subject of intrigue and speculation since the 1880s.

The phenomenon gained national attention when it was featured in the August 1969 issue of National Geographic Magazine (Volume 136, No. 2). The origin story of these lights is as intriguing as the lights themselves. It is said that they were first noticed by a group of miners who were using the cemetery as a shortcut. The miners, having lost their way, witnessed the mysterious lights emerging within the cemetery. This sighting marked the beginning of the cemetery's association with these enigmatic lights, drawing increased interest and visitors.

Some locals attribute these lights to a natural phenomenon known as "Wildfire," typically seen in swamps and damp locations. However, the true nature of these lights remains a topic of debate and fascination.

Adding to the intrigue, researcher and writer Karen Stollznow conducted an investigation into these lights. Stollznow proposed several natural explanations, such as a specific retinal receptor sensitive to blue light or the phenomenon of phosphenes—brief, eye-induced flashes of light. These scientific explanations suggest that the lights' apparent movements - "dancing," "floating," or "darting" - could be illusions created by the eye and mind, especially after prolonged periods of anticipation and observation in the dark. Stollznow's research indicates that the experience of witnessing these lights is highly subjective, shaped largely by the observer's expectations and state of mind.

Silver Cliff Cemetery thus stands not just as a resting place for the departed, but also as a crossroads where history, culture, and the unexplained converge, making it a fascinating destination for those interested in the mysteries of the natural world and the narratives of the past.

The Spooklight

The Spooklight phenomenon, also known by various names like the Hornet Spooklight, Hollis Light, and Joplin Spook Light, presents a mysterious and captivating tale that has intrigued locals and visitors alike. This enigmatic ghost light, shimmering on the border of southwestern Missouri and northeastern Oklahoma, near the quaint town of Hornet, Missouri, has been a subject of speculation and folklore for decades. The light appears along a stretch near the historic Route 66, just south of Quapaw, Oklahoma.

Interestingly, this phenomenon occurs in a specific geographical alignment. An east-west stretch of Route 66 aligns perfectly with a farm road known as E 50, or "Spooklight Road", located approximately ten miles east, across the Spring River. The intrigue deepens when viewed from higher vantage points along E 50. From these points, the headlights of cars traveling east on Route 66 manifest as ghostly lights in the distance, giving birth to the legend of the Spooklight.

The first documented recognition of this optical illusion was by AB MacDonald in a January 1936 issue of the Kansas City Star. Since then, the Spooklight has been the subject of numerous experiments and investigations. These have included the use of fireworks, spotlights, and even strategically placed car headlights along Route 66, all observed from Spooklight Road. These experiments have successfully replicated the mysterious light, demystifying the legend to some extent.

Notable investigations include Thomas Sheard's study in 1946, a thorough investigation by a Kansas City group in 1955, research by Robert Gannon in 1965, and a more recent inquiry by Allen Rice and his team of "Boomers" sleuths in 2015. Each of these studies has contributed to the understanding of the Spooklight, transforming it from a supernatural enigma to a fascinating case of optical illusion and geographical alignment. Despite these explanations, the Spooklight continues to capture the imagination of locals and visitors, remaining a charming and mysterious part of the region's folklore.

Like many other ghost lights around the world, the Spooklight has inspired a rich tapestry of folklore and legend, with storytellers often suggesting its existence long before the advent of automobiles. However, these claims lack verification from printed sources. Journalist Paul W. Johns, through his meticulous research, concluded that there is no record of the Spooklight in print before 1926, coinciding with the designation of that section of Route 66.

The 1960s saw the emergence of a Spooklight museum at the eastern end of E 50. However, in a Popular Mechanics article, Robert Gannon described it as a lackluster "tourist trap." The museum featured a three-inch telescope for viewing the light at a fee of 25 cents. Yet, the setup was indoors, looking through a small hole in the wall, which significantly reduced its resolving power. The proprietor justified this setup as necessary for protecting the telescope from rain. Gannon, armed with a comparable telescope, noted that while the naked eye saw a single light, his telescope distinctly revealed it as car headlights, always appearing in pairs.

Local businesses and chambers of commerce have historically capitalized on the Spooklight's allure for generating tourism revenue. In 1969, the Missouri Chamber of Commerce released a press statement, falsely claiming that scientists, despite using various technical devices, couldn't determine the light's origin. Additionally, the Joplin Chamber of Commerce published "The Tri-State Spook Light" visitor guidebook in 1955, and the Neosho Chamber of Commerce released its tourist booklet in 1963. The Missouri Division of the

U.S. Brewers Foundation even ran newspaper ads in the 1950s, promoting the Spooklight in anticipation of boosting beer sales to tourists.

Online lore about the Spooklight often includes the claim that a manuscript titled "The Ozark Spook Light," supposedly written by Foster Young in the 1880s, counters the headlight explanation. However, no evidence supports the existence of either the document or the author, leaving these stories in the realm of unverified legend and folklore.

II

Around the World

Kitsunebi

Kitsunebi, or "fox fires," hold a fascinating place in Japanese folklore, intriguing and mystifying people with their ethereal presence. These ghostly lights, renowned throughout Japan except in Okinawa Prefecture, are also known by various other names such as "hitobosu," "hitomoshi," and "rinka."

Kimimori Sarashina, a dedicated researcher of local stories, provides a captivating summary of these enigmatic lights. Imagine a scene where, in the absence of any obvious source of fire, mysterious flames suddenly appear. These lights, resembling paper lanterns or torches, form a line and flicker unpredictably. In a baffling display, extinguished fires might reappear elsewhere, creating a chase where the source continually vanishes.

These lights are not random occurrences; they favor specific times of the year. Most commonly witnessed between spring and autumn, their appearances are particularly frequent in the sweltering heat of summer, often emerging during cloudy weather or when the atmosphere is ripe with change.

The number of these lights can be astonishing, ranging from ten to several hundred in a line. Just when one thinks their number has increased, they vanish, only to multiply again. In Nagano Prefecture, for example, numerous lights resembling paper lanterns line up and flicker in a mesmerizing dance.

The length of these mysterious lines can be impressive, stretching up to 600 meters. While the predominant colors of these fires are red or orange, there have been numerous accounts of witnesses encountering the rare sight of blue flames. These kitsunebi are not just a mere natural phenomenon; they are a bridge to the mystical, a glimpse into a world that blurs the line between reality and the supernatural.

In Tonami, Toyama Prefecture, these enigmatic lights are said to appear on remote hillsides, far from any roads or human presence. Contrasting this, in Monzen, formerly part of Fugeshi District in Ishikawa Prefecture (now part of Wajima), legends tell of kitsunebi that eerily follow humans wherever they go. This aligns with the broader Japanese folklore narrative where foxes are often depicted as cunning beings that trick humans, with kitsunebi lighting up pathless areas to lead people astray.

In Iida, Nagano Prefecture, locals believed that one could dispel these misleading lights with a simple yet bold act of kicking them away with one's feet. Meanwhile, in Izumo Province, now known as Shimane Prefecture, encounters with kitsunebi were said to bring fevers, reinforcing the belief that these lights were akin to ikiaigami, divine spirits capable of bringing misfortune to those unprepared for such encounters.

Adding to the tapestry of stories is a legend from Nagano, where a lord and his vassal, in search of a location to build a castle, were guided at night by a white fox whose light illuminated their path, leading them to an ideal site. This tale highlights the more benevolent aspects of fox spirits in Japanese folklore.

Kitsunebi have also inspired literary works, such as the haikus composed by Masaoka Shiki, who reflected on their beauty in the context of winter landscapes. While commonly associated with winter, there are accounts of kitsunebi appearing in the heat of summer or the transitional season of autumn, demonstrating their unpredictable nature.

The relationship between kitsunebi and another mysterious light phenomenon known as onibi is a subject of debate. While some theories suggest that kitsunebi is another term for onibi, typically, they are considered distinct entities in Japanese folklore.

The Ōji Inari Shrine in Ōji, Kita, Tokyo, stands as a prominent and revered site in Japanese folklore. Known as the head sanctuary of Inari Ōkami, it is not just a spiritual center but also a legendary hub for witnessing kitsunebi, the mystical fox fires.

In the past, the area surrounding Ōji Inari was predominantly rural, a landscape punctuated by a grand enoki tree standing tall by the roadside. This tree was not just a natural landmark but a focal point of enchanting folklore. On the eve of Ōmisoka, the last night of the year, it was believed that foxes from across the entire Kanto region, referred to as Kanhasshū, would converge beneath this tree. In an extraordinary spectacle, these foxes were said to don uniforms, call out their ranks, and embark on a ceremonial procession to the palace of Ōji Inari.

The kitsunebi, visible during this unique gathering, presented a breathtaking sight. Local farmers and residents held the belief that the number of these ethereal lights could foretell the bounty of the coming year's harvest. Thus, on Ōmisoka night, they would gather to count these flickering flames, hoping to predict whether the next year would bring fortune or famine.

The enoki tree, central to this tradition, became so intertwined with these events that it earned the nickname "shōzoku enoki," which translates to "costume enoki." This site gained such renown that it even inspired the famous ukiyo-e artist Hiroshige, who immortalized it in his series "One Hundred Famous Views of Edo." Sadly, the original enoki tree withered away during the Meiji period, but its legacy endures. A small shrine, the Shōzoku Inari Jinja, stands as a testament to this history, located near what used to be the second Ōji tram stop, now at the "horibun" intersection.

The area, once known as Enokimachi or "enoki town," has not forgotten its rich cultural heritage. In 1993, in recognition of this legendary tradition and as part of a broader development plan, the "Ōji Kitsune's Procession" event was inaugurated. Held annually on the evening of Ōmisoka, this event serves as a modern-day homage to the mystical procession of the past.

The manifestation of kitsunebi, or "fox fires," varies intriguingly across different regions, each with its unique cultural interpretation. In the Dewa Province, encompassing parts of modern-day Yamagata Prefecture and Akita Prefecture, these ethereal lights are referred to as "kitsune taimatsu," literally translating to "fox torch." Here, the kitsunebi are romantically linked to the folklore of fox marriages. It is believed that these lights serve as torches illuminating the path for a fox wedding procession, symbolizing a joyous and auspicious event, and are thus regarded as a good omen.

Moving to Bizen in Okayama Prefecture and the Tottori Prefecture, the local interpretation of these atmospheric ghost lights takes on a different name and nature. Known as "chūko," which translates to "mid-air fox," these lights possess a distinct characteristic from the typical kitsunebi. In Toyohara village, situated in the Oku District of Okayama, a local legend suggests that an aged fox transformed into a chūko, a low-altitude floating light. This transformation is a vivid illustration of the fox's mystical abilities in folklore.

On Ryūgūjima, in Tamatsu village of the same district, the appearance of chūko is connected with meteorological phenomena. These lights, resembling the size of paper lanterns, emerge at night and are seen as harbingers of impending rain. In an almost magical display, these chūko sometimes descend to the earth, lighting up the surroundings with an otherworldly glow before vanishing without a trace, leaving no evidence of their ephemeral visit.

Enryō Inoue, a renowned yōkai (supernatural creature) researcher from the Meiji period, contributed significantly to the classification of these mysterious lights. He used the kanji characters to denote chūko, establishing a categorization based on their altitude. According to his classification, "tenko" (), meaning "heavenly fox," are those that fly high, while chūko are identified as those that float closer to the ground.

One common thread in these legends involves the depiction of foxes whose breath emits a mysterious glow. Another captivating aspect is the belief that foxes could strike their tails against the ground to ignite fires. Perhaps most intriguing is the notion of the "kitsunebi-tama" or "kitsunebi ball," a luminescent orb associated with these ethereal creatures, adding to their supernatural aura.

In the Shokoku Rijidan, an essay from the Kanpō period, there is a fascinating account from the early Genroku years. Fishermen, it was said, would sometimes capture kitsunebi within their nets, only to discover a kitsunebi-tama entangled among their catch. This object was prized for its unique property of providing illumination only at night, remaining inert during daylight hours.

The Genroku period also brought forth the Honchō Shokkan, a book on herbalism, which contains an interesting assertion regarding foxes. It suggests that these creatures used withered trees on the ground to create fires. In English, "fox fire" translates to "kitsunebi" in Japanese, but this term does not refer to the animal fox. Instead, it signifies something withered or decayed. This ties back to the concept of "fox fire" being related to the light from hypha and mushroom roots that cling to decaying trees. The Honchō Shokkan's references could be interpreted as alluding to the luminescence from hypha on these withered trees.

Moreover, the Honchō Shokkan and other literary works like the Meiwa period's Kunmō Tenchiben by Takai Ranzan and the late Edo Period's Shōzan Chomon Kishū by Miyoshi Shōzan also mention foxes producing light using bones such as human skulls or horse bones. In the Shinshū Hyaku Monogatari, a collection of strange tales from Nagano Prefecture, there's a story where a person approaching a kitsunebi would encounter a fox holding glowing human bones in its mouth, emitting a turquoise light.

These accounts led researchers like Enryō Inoue to theorize that phosphorous light from bones might be connected to the phenomenon of kitsunebi. Phosphorus, known to spontaneously combust at temperatures above 60 degrees Celsius, adds a scientific angle to the folklore, linking the fox's identity to phosphorus-induced light. However, this theory struggles to explain sightings of kitsunebi visible from kilometers away, which would be unlikely for light sources as faint as hypha or phosphorus.

Yoshiharu Tsunda, a folklorist in 1977, proposed a more grounded explanation, suggesting that most kitsunebi sightings could be attributed to significant light refractions common in alluvial fans between mountainous and plain regions. While some have hypothesized other natural phenomena like the spontaneous combustion of petroleum or ball lightning as possible explanations, many mysteries of kitsunebi remain unexplained, leaving room for the imagination and keeping the enchantment of these tales alive in Japanese folklore.

Shiranui

The mysterious and enchanting phenomenon of "shiranui," also known as "phantom fires," unfolds in the open waters, several kilometers from the shore. Initially, observers may notice one or two flames, locally referred to as "oyabi" (, meaning "parent-fire"), gently flickering in the distance. These ethereal flames undergo a mesmerizing transformation, dividing and multiplying to the left and right, eventually forming an awe-inspiring spectacle of several hundred to even thousands of individual flames. These ghostly fires stretch impressively across four to eight kilometers, creating a breathtaking sight in the dead of night.

It is widely believed that the shiranui are most prominent and numerous at the lowest tides, especially around two hours before and after 3:00 AM. Interestingly, these flames are said to be elusive and invisible from the water's surface, yet they become visible at elevations up to ten meters above sea level. Those who dare to approach or chase these flames will find them receding further into the distance, as if playing a tantalizing game of hide and seek.

Historically, the shiranui were shrouded in myth and reverence. Local fishing communities, once believing them to be the glowing lamps of the Dragon God, would observe strict fishing prohibitions on days when the shiranui made their mysterious appearance. This deep-rooted superstition reflects the cultural significance and mystical aura surrounding these phantom fires.

Intriguingly, ancient texts like the Nihon Shoki, the Hizen no Kuni Fudoki, and the Higo no Kuni Fudoki, recount how Emperor Keiko, during his conquest of Kumamoto in southern Kyushu, utilized the shiranui as navigational beacons. On his quest to expand the Yamato Ōken, the Emperor encountered these unexplained, moving spots of fire in the Ariake Sea and the Yatsushiro Sea, also known as the Shiranui Sea. When he inquired about these mysterious fires with the local elite, or gozoku, they professed ignorance of such phenomena. Consequently, these enigmatic flames were named "shiranu hi" (meaning "unknown fire"), which linguistically evolved into "shiranui." This led to the naming of the surrounding region as Hi Province (or , Hi no kuni), translated as "Land of Fire" or "Kingdom of Fire," now recognized as Kumamoto Prefecture. .

In the captivating and historically rich pages of Nakashima Hiroashi's 1835 tome "On the Shiranui," a detailed and enthralling account of the elusive and mystical phenomenon of Shiranui, or the "unknown fire," is unveiled. This extraordinary natural marvel, a source of both awe and intrigue, has beckoned countless individuals from far-flung regions to the Gulf of Shimabara, near the Ariake Sea in Japan, eager to witness its enigmatic beauty.

Hiroashi, through his meticulous observations, reveals that the Shiranui manifests itself as ethereal fires, visible from a distance of 8 to 12 kilometers off the seashore. Although the exact distance remains shrouded in the ambiguity of the dark night, the initial appearance of one or two fiery spots, known as oya-dama or "father fire," marks the beginning of a mesmerizing display. As the night deepens, these solitary fires multiply and extend, creating a dynamic spectacle of flickering lights that resemble stars sprinkled across the night sky. This stunning array of lights ebbs and flows in intensity until the break of dawn, when they gradually fade away.

Hiroashi notes an intriguing aspect of Shiranui's behavior: the phenomenon is conspicuously absent during inclement weather, such as rain or strong winds. This curious characteristic adds to the mystique and unpredictability of the Shiranui, making its appearance all the more sought after and cherished.

The tradition of Shiranui watching has evolved into a cultural event, with many people ascending to vantage points on nearby mountains. There, amidst the backdrop of nature's nocturnal canvas, they partake in the communal joy of sake drinking, reveling in the shared experience of witnessing this mysterious natural wonder.

Furthermore, the Shiranui is not a constant spectacle but is known to occur around specific times of the year, notably around September 30th and February 24th. During these periods, from midnight until the early hours of dawn, the Shiranui takes on various breathtaking forms. Sometimes, it appears as a colossal ball of fire, soaring up to 60 feet above the sea's surface. On other occasions, it transforms into a mesmerizing procession of pale red, fiery globes that gracefully drift along the tide.

During the Taishō era, the mesmerizing phenomenon of Shiranui, a mysterious display of phantom lights off the coast of Japan, captivated not only the general public but also piqued the interest of scholars and journalists. This enigmatic occurrence, often likened to a natural light show, sparked numerous scientific attempts to unravel its secrets, leading to various theories and extensive investigations.

In a notable endeavor to demystify Shiranui, a large-scale investigation was launched in 1916. This ambitious project, involving two ships and a crew of over 50 people, set sail into the depths of the Ariake Sea. Despite the scale and dedication of this expedition, the results were inconclusive and muddled with conflicting data, leaving the scientific community without a definitive explanation for the phenomenon.

One prominent theory, emerging during the Shōwa era, was proposed by Machika Miyanishi, a distinguished professor at both Kumamoto Higher Technical School and Hiroshima Higher Technical School. Miyanishi, who dedicated extensive research to this phenomenon, authoring eight scholarly papers including two in English, suggested a fascinating interplay of natural and human factors. He posited that the appearance of Shiranui coincided with the peak sea temperatures of the year and a significant tidal retreat, exposing mudflats and causing sudden radiative cooling. This natural setting, combined with the unique land formations at the Yatsushiro Sea and the Ariake Sea, could refract the lights from boats venturing out for tidal fishing. Miyanishi's theory, which also speculated that the lights were traditional fires used to lure fish at night (known as isaribi), gained recognition for its plausible explanation of the mysterious flames as a result of optical illusions and natural factors.

In another scholarly pursuit, the renowned meteorologist Sakuhei Fujiwhara delved into the realm of atmospheric light phenomena with his publication in 1933. Despite his expertise, Fujiwhara confessed his uncertainty regarding the cause of Shiranui. He also raised an intriguing possibility, suggesting that the phenomenon might be influenced by the presence of noctiluca, a type of bioluminescent organism. Additionally, he speculated on the role of local entrepreneurs, who, capitalizing on the public's fascination with Shiranui, may have contributed to the spectacle, enhancing its allure for eager sightseers.

Tairi Yamashita, a distinguished Professor at Kumamoto University, embarked on an extensive journey to unravel the mysteries of Shiranui, employing modern instruments and harnessing the collaborative efforts of his students. Yamashita's research led to a significant breakthrough, concluding that Shiranui is an intricate optical phenomenon. This spectacle results from light traversing through complex distributions of air clumps, each varying in temperature, and undergoing refraction. He posited that the source of these ethereal lights could be traced back to mundane origins such as domestic fires and lights used by fishermen to lure their catch. According to his findings, similar optical phenomena, akin to road mirages, general mirages, and heat shimmers, can manifest under comparable conditions in various locations and times.

Complementing these scientific endeavors, Nobuyuki Marume, in his captivating collection of essays titled "Shiranui," presented a rich visual documentation of the phenomenon. His work, "Changes of Shiranui with the Passage of Time from Shiranui Town towards the Direction of Amura," features an array of photographs capturing the transformative nature of Shiranui over time, offering a visual testament to its ever-changing beauty.

Machika Miyanishi, a professor at both Kumamoto Higher Technical School and Hiroshima Higher Technical School, also dedicated extensive research to Shiranui, producing eight papers, including two in English, on the subject. Miyanishi's hypothesis centered around the light from fishing boats as the primary source of Shiranui. He suggested that on particular days, the formation of large tidelands results in air lumps of differing temperatures, which, through the vortex movement of these air masses, distort and fluctuate the light, creating the illusion of Shiranui. Similarly, Tairi Yamashita's studies corroborated the idea of a dynamic interplay between hot and cold air clumps, leading to fluctuating and often splitting light sources. These irregular and dramatic changes, according to his research, could lead to the manifestation of Shiranui under the right conditions, even in different seasons.

In recent times, however, the once prevalent Shiranui has become a rarity. The transformation of the coastal landscape, marked by the filling of tidal flats, the omnipresence of electric lights piercing the night's darkness, and the unfortunate pollution of seawater, has greatly diminished the chances of witnessing this natural marvel. The decline of Shiranui serves as a poignant reminder of the delicate balance between nature and human activity, highlighting the impact of environmental changes on natural phenomena that once captivated generations.

Kitsune no Yomeiri

"The Kitsune no Yomeiri" or "The Fox's Wedding" is a captivating and mysterious term steeped in the folklore and cultural heritage of Japan, particularly prevalent in the regions of Honshu, Shikoku, and Kyushu. This intriguing phrase encapsulates a range of natural phenomena and folk beliefs tied to the supernatural, weaving a tapestry of myth and reality. At the heart of these phenomena is the enigmatic and often playful figure of the fox, or 'kitsune', a central character in numerous Japanese legends known for their cunning and trickster nature.

One of the most fascinating interpretations of the "Kitsune no Yomeiri" relates to the phenomenon of atmospheric ghost lights. These ephemeral lights resemble paper lanterns from a wedding procession, floating ethereally through the darkness of night, creating an otherworldly and mesmerizing spectacle. They evoke a sense of mystery and enchantment, as if one is witnessing a secret ceremony from a hidden world.

Another aspect of this folklore is its association with sunshowers, a rare and beautiful weather phenomenon where rain falls even as the sun shines. This natural marvel has often been likened to the mystical wedding processions of foxes, further entwining natural events with folklore.

The "Kitsune no Yomeiri" also finds its place in classical Japanese 'kaidan' (ghost stories), essays, and legends, where various phenomena resembling wedding processions are narrated with a blend of awe and mystery. These stories often feature foxes as central figures, highlighting their significant role in Japanese mythology.

In a remarkable testament to the cultural impact of this folklore, various Shinto rituals and festive rites have been developed in different parts of Japan, celebrating the enigmatic and intriguing nature of the "Kitsune no Yomeiri." These practices not only honor the folklore but also reflect the deep connection between the Japanese people and their rich mythological heritage.

An example of the historical documentation of this phenomenon can be found in a topography book of the Echigo Province (now Niigata Prefecture), from the Hōreki period. The "Echigo Nayose" contains a detailed description of the atmospheric ghost lights associated with the "kitsune no yomeiri." The account describes how, on exceptionally quiet nights, flames resembling paper lanterns or torches can be seen extending into the distance, far beyond what one would expect. These sightings, though rare, have been particularly noted in the Kanbara district, where they are referred to by the youth as "the wedding of foxes."

In regions stretching from Niigata Prefecture to Tokyo, and from Yamanashi to Tottori Prefecture, the appearance of atmospheric ghost lights, known as 'kitsunebi', creates a mystical spectacle. These lights, often extending nearly 4 kilometers, evoke the image of a grand and otherworldly procession. In locales like Nakakubiki District, Uonuma, Akita Prefecture, and several others, these mesmerizing lights are referred to as "kitsune no yomeiri," a term that conjures the image of a fox's wedding procession. The sight is so captivating and surreal that it's as if the quiet countryside is momentarily transformed into a stage for a mystical event.

Interestingly, the phenomenon is known by different names in other areas, each with its own unique twist on the legend. For example, in Sōka, Saitama Prefecture, and Noto, Ishikawa Prefecture, it's called "kitsune no yometori", translating to "the fox's wife-taking." Meanwhile, in Numazu, Shizuoka

Prefecture, it's referred to as "kitsune no shūgen", or "the fox's celebration." These variations in names across different regions highlight the rich diversity of folklore interpretations in Japan.

The connection between these atmospheric ghost lights and traditional Japanese wedding practices is particularly fascinating. Until the middle of the Showa period, it was customary for weddings to be held in the evening, often marked by a procession of paper lanterns. This historical context adds depth to the folklore, as the ghost lights, resembling a line of lanterns and torches, are reminiscent of these traditional wedding processions.

There are intriguing theories as to why foxes are seen as the protagonists in these mysterious events. One theory suggests that the lights, which appear to signal a wedding but vanish upon closer inspection, are an elaborate illusion crafted by playful foxes. This idea plays into the broader cultural portrayal of foxes in Japanese folklore as cunning tricksters, capable of creating enchanting yet deceptive scenes. The illusion is so convincing that from afar, the lights resemble a procession of paper lanterns, but upon approach, they disappear, much like the cunning of a fox leading one on a merry chase.

In Toyoshima, once part of old Edo and now nestled in the Toshima and Kita wards of Tokyo, the phenomenon of atmospheric ghost lights, quivering and dancing in the dark, is revered as "kitsune no yomeiri." This spectacle is so integral to local lore that it is counted among the "seven mysteries of Toshima," a collection of enigmatic tales and legends that have been passed down through generations in the village. The image of these ghostly lights, wavering mysteriously in the night, has long captured the imagination of the local people, imbuing the area with a sense of wonder and mystique.

In the lush region of Kirinzan, Niigata Prefecture, known for its abundant fox population, there are tales of nocturnal wedding processions illuminated by paper lanterns, believed to be the celebrations of foxes. This belief ties in closely with local agriculture; it is said that the abundance of these atmospheric ghost lights foretells a year of bountiful harvests, while their scarcity predicts a year of poor crops. This connection between the natural, supernatural, and agricultural practices highlights the deep interlinking of folklore and daily life in rural Japan.

The legends surrounding "kitsune no yomeiri" go beyond mere sightings of ghost lights in some areas. In Gyōda, Saitama Prefecture, it is believed that these mystical fox weddings frequently occur near the Kasuga Shrine in Tanigou. Intriguingly, after such reported events, traces of fox feces are sometimes found along the road, as if to confirm the passage of a mystical procession. Similarly, in Horado, Mugi District, Gifu Prefecture, residents report not just the sighting of atmospheric ghost lights, but also the eerie sound of bamboo burning and tearing over several days, with no physical trace left behind to explain these auditory phenomena.

In stark contrast to these celebratory interpretations, the Tokushima Prefecture views these occurrences through a more somber lens. Here, the atmospheric ghost lights are not seen as heralds of fox weddings but as signals of fox funerals, ominously predicting the impending death of someone in the community. This variation showcases the diverse cultural perspectives within Japan on the same natural phenomenon.

In the wide expanse of the Kantō, Chūbu, Kansai, Chūgoku, Shikoku, and Kyushu regions, the phenomenon of sunshowers – where rain falls despite the presence of sunshine – is commonly referred to as "kitsune no yomeiri." This intriguing association between weather and folklore is a testament to the rich imagination and cultural depth of the Japanese people.

The naming of this phenomenon varies intriguingly across different locales. In the Nanbu Region of Aomori Prefecture, it is known as "kitsune no yometori", translating to "the fox's wife-taking." In Serizawa, Chigasaki, Kanagawa Prefecture, and the mountainous areas of Oe District, Tokushima Prefecture, it is called "kitsune-ame" (fox rain). The eastern Isumi District in Chiba Prefecture refers to it as "kitsune no shūgen". Meanwhile, in the Higashi-Katsushika District, Chiba Prefecture, it's known as "kitsune no yometori ame" (the fox's wife-taking rain), a name steeped in the region's agricultural heritage where wives were traditionally valued for their labor and seen as essential to the family's prosperity.

The reasons behind linking sunshowers with foxes, wives, and weddings are as diverse and imaginative as the regions themselves. One explanation suggests that the paradoxical nature of a sunshower, where rain falls under clear skies, mirrors the trickery and illusion often attributed to foxes in folklore. People felt as if they were witnessing an improbable, almost magical event. Another theory posits that foxes chose sunshowers for their weddings, creating a romantic and mystical imagery. In some areas, it was believed that foxes caused rain to fall at the base of mountains to prevent humans from witnessing their sacred nuptials. Another interesting interpretation relates to the emotional complexity of weddings, where joy (sunshine) and tears (rain) often coexist, symbolizing the bittersweet nature of such life events.

Furthermore, the association between a fox's wedding and specific weather patterns varies greatly by region. In Kumamoto Prefecture, a rainbow's appearance signals a fox's wedding, while in Aichi Prefecture, it is the fall of graupel, a type of precipitation similar to hail. These regional variations in the legend of "Kitsune no Yomeiri" reflect the diverse climatic conditions and cultural landscapes of Japan, illustrating how folklore adapts and thrives in different environments.

In historical literature, there are numerous accounts of actual sightings and encounters with these mystical fox weddings. For instance, the "Konjaku Yōdan Shū," a collection of supernatural tales from the Kan'ei period, recounts a strange wedding procession in Takemachi, in the Honjo area of Edo. Similarly, the "Edo Chirihiroi," a compendium of Edo anecdotes, describes a fox wedding witnessed at the Hacchō canal. Additionally, "Kaidan Oi no Tsue," a collection of ghost stories from the Kansei period, narrates an encounter with a fox wedding in the village of Kanda, Kōzuke (now part of Gunma Prefecture). These tales, documented in historical texts, lend an air of authenticity and depth to the folklore, bridging the gap between legend and reality.

Across Japan, stories of marriages between foxes revealed to humans have been shared and cherished. In Sōka, Saitama Prefecture, a poignant folk legend from the Sengoku period tells of a woman who, having promised to marry her lover, tragically dies from an illness. Moved by the sadness of the unfulfilled promise, foxes are said to have emulated a wedding procession near the woman's grave. Similarly, a tale from Shinano Province speaks of an old man who once aided a young fox. In gratitude, the fox, upon reaching maturity, invites the old man to join its magnificent wedding procession. These stories not only reflect the deep connection between humans and the mystical world of foxes but also highlight the cultural importance of empathy and gratitude.

The settings for these fox weddings are as varied and mystical as the stories themselves. Often, these nuptials are intertwined with natural and super-natural phenomena. In tales of daytime weddings, sunshowers serve as a magical backdrop, symbolizing the union of joy and sorrow, sunshine and rain. Nighttime weddings, on the other hand, are frequently set amidst the ethereal glow of atmospheric ghost lights, creating an ambiance of mystery and enchantment. These natural occurrences function like stage settings, transforming ordinary landscapes into realms of fantasy and wonder.

In Fukushima Prefecture, there is a unique and somewhat whimsical ritual believed to grant one the ability to witness a fox's wedding. On the evening of October 10th, according to the lunisolar calendar, it is said that if a person places a suribachi (a traditional Japanese mortar) on their head, sticks a wooden pestle in their waistband, and stands under a date plum tree, they can see the elusive fox wedding procession. Similarly, in Aichi Prefecture, another curious method is described: spitting into a well, interlocking one's fingers, and peering through the gap between them is believed to reveal the mystical sight of a fox wedding.

The lore of "Kitsune no Yomeiri" also delves into tales of unions between humans and foxes, transcending the boundaries between species. A notable example is the story of Abe no Seimei, a legendary Heian period onmyoji (a practitioner of Japanese esoteric cosmology), whose birth is attributed to the union of a human and a fox spirit, as depicted in the tale of Kuzunoha. This story has become so ingrained in Japanese culture that it was adapted into ningyō jōruri, a form of traditional Japanese puppet theater.

Furthermore, similar narratives are found in historical texts like the "Nihonkoku Genpō Zen'aku Ryōiki" and the "Tonegawa Zushi," a topographical book published in 1857. The latter recounts a story involving a commander named , likened to the legendary Chinese strategist Zhuge Liang, and the town of Onabake ("shapeshift into woman") in Ushiku, Ibaraki Prefecture, which is said to have derived its name from such a tale. The Onabake Jinja in Ryūgasaki, in the same prefecture, even deifies a fox, further cementing the legend's place in local culture.

Intriguingly, these tales often involve shape-shifting foxes assuming human forms to interact with people. For instance, in the "Konjaku Monogatarishū" and the "Honchō Koji Innen Shū," published in 1689, as well as the "Tamahahaki," published in 1696, there are stories of foxes transforming into human wives. Conversely, in the "Tonoigusa," a collection of ghost stories published in 1677, there is a tale of a male fox who falls in love with a human woman, takes on the guise of her husband, and even fathers children with her.

These stories showcase the creative depth of Japanese folklore and depict foxes as mystical beings that blur the line between the natural and supernatural. They reveal a world where foxes are seen as enchanting entities with transformative powers, significantly impacting human lives.

Onibi

The Onibi, or "Demon Fire," is an entrancing and mysterious phenomenon deeply rooted in Japanese folklore. These spectral lights, resembling ghostly flames, are believed to originate from the souls of both humans and animals. Folklore often depicts them as manifestations of vengeful spirits transformed into fiery apparitions. These ethereal entities also find their counterparts in Western folklore, where they are akin to the "will-o'-wisp" or "jack-o'-lantern," terms that are occasionally used as equivalents to Onibi in Japanese translations.

Historically, the Wakan Sansai Zue, a Japanese encyclopedia from the Edo period, provides an intriguing description of the Onibi. It portrays them as blue, pine torch-like lights that congregate and pose a supernatural threat to humans, reportedly sapping away their life essence upon close contact. This vivid depiction is further embellished by illustrations from the same period, suggesting that the Onibi vary significantly in size - from a mere few centimeters to as large as 20 or 30 centimeters in diameter. They are also depicted as floating eerily one to two meters above the ground, adding to their ghostly presence.

One particularly fascinating anecdote from the Edo period, recorded in the essay "Mimibukuro" by Yasumori Negishi, recounts an incident involving an Onibi above Hakone Mountain. This Onibi intriguingly split into two separate entities, then reconverged, only to split again multiple times, displaying a mesmerizing and unpredictable nature.

In contemporary times, numerous theories have been proposed to explain the appearance and characteristics of these enigmatic lights.

Regarding their appearance, while they are predominantly blue, as earlier accounts suggest, variations exist in hues ranging from bluish-white, red, to yellow. The size of these ghostly flames is equally variable, with some being as small as a candle flame, while others can grow to the size of a human or even span several meters.

The number of Onibi sightings can vary dramatically. Sometimes, only a solitary light or a pair is observed, while at other times, as many as 20 to 30 can be seen together. There are even accounts of countless Onibi flickering throughout the night, only to vanish with the dawn.

These spectral flames are most frequently sighted from spring through summer, often emerging on rainy days. Their preferred haunts include watery areas like wetlands, forests, prairies, and graveyards, locations that are typically surrounded by natural landscapes. However, on rare occasions, they have been seen in urban areas.

An intriguing aspect of the Onibi is their heat - or lack thereof. Some reports indicate that touching them yields no sensation of heat, akin to a benign phantom flame. In contrast, other accounts suggest they possess the ability to burn like real fire, potentially causing harm or destruction.

These mysterious, ethereal lights are not just limited to the commonly known Onibi; there are numerous other types, each with its own unique characteristics and legends. Among these are the Shiranui, Koemonbi, Janjanbi, and Tenka. There's also the intriguing Kitsunebi, often debated whether it's a type of Onibi or a distinct phenomenon altogether.

One captivating variant is the Asobibi, also known as "play fire." This type of Onibi is known to make appearances in specific locations such as below castles and above seas in Kōchi Prefecture, and Mitani Mountain. Witnesses describe its behavior as playful and elusive; it seems to be close one moment and then suddenly flies far away. It's known for splitting into several parts only to merge back into one. Remarkably, Asobibi is believed to be harmless to humans.

In the Watarai District of Mie Prefecture, Onibi are referred to as Igebo, highlighting regional variations in naming these mysterious lights. Another intriguing type is the Inka, or "shadow fire," which is said to appear alongside ghosts or yōkai, adding an extra layer of mystery to supernatural encounters.

The Kazedama, or "wind ball," is a unique Onibi from Ibigawa in Gifu Prefecture. It manifests as a spherical ball of fire, particularly during storms, and is described as being as large as a personal tray. This Onibi became notably prominent during the typhoon in Meiji 30 (1897), where it appeared from the mountains and floated in the air multiple times.

Sarakazoe, or "count plate," is an Onibi immortalized in the Konjaku Gazu Zoku Hyakki by Sekien Toriyama. It's associated with the famous ghost story of Banchō Sarayashiki, where the spirit of Okiku is depicted as an Inka emerging from a well and counting plates.

The Sōgenbi, another Onibi type from Kyoto, has a more somber backstory. Featured in Sekien Toriyama's Gazu Hyakki Yagyō, it is said to be the spirit of a monk who committed theft and received Buddhist punishment, transforming into an Onibi. The tormented face of the priest is said to be visible within the flames.

In Okinawa Prefecture, there's the Hidama, or "fire spirit." This Onibi is usually found in the kitchen behind charcoal extinguishers but is believed to transform into a bird-like shape, causing fires as it flies around.

The Wataribisyaku, from Chii village in Kyoto Prefecture, appears as a bluish-white ball of fire, floating lightly in the air. Its appearance is likened to a hishaku (ladle), not because it resembles the tool, but due to the long, thin tail it seems to pull behind it, metaphorically compared to a ladle.

Lastly, the Kitsunebi, or "fox fire," is enveloped in various legends. One theory suggests that the glow comes from a bone held in a fox's mouth, while others explain it as a light refraction near river beds. Sometimes considered a type of Onibi, Kitsunebi continues to intrigue and inspire.

The term Onibi itself translates to "demon fire," yet these phenomena are far more complex and varied than simple flames of combustion. The inconsistencies in eyewitness accounts suggest that Onibi may encompass a range of luminous events rather than a single type of occurrence. Notably, their frequent appearance during rainy days further distances them from typical fires.

The concept of Onibi is not unique to Japan. Ancient Chinese texts, dating back to the BC era, spoke of lights born from the blood of humans and animals, believed to be a mix of phosphorus and "oni fire" (Onibi). Interestingly, the character in these texts referred to a range of luminescent phenomena, including the glow of fireflies and triboelectricity, rather than the chemical element phosphorus as it is understood today.

In Japan, the "Wakan Sansai Zue," a Japanese encyclopedia, offers a poignant explanation, suggesting that Onibi are the transformed spirits of humans, horses, and cattle that died in battle, their blood staining the ground. This interpretation indicates a deep connection between Onibi and the spiritual world, a common theme in many cultures' folklore.

The 19th century saw further evolution in the understanding of Onibi. Literary works like Shūkichi Arai's "Fushigi Benmō" proposed that Onibi were born from the phosphorus in human corpses. This view, influenced by the discovery of phosphorus in human bodies by biologist Sankyō Kanda in 1696, gained traction and was widely accepted well into the 20th century.

However, modern science offers alternative theories. Some suggest that Onibi could be the result of spontaneous combustion of phosphine, methane produced by decaying corpses, or even hydrogen sulfide emanating from decay. Others propose that these mysterious fires are a type of plasma, akin to Saint Elmo's fire, especially given their propensity to appear during rainy conditions. Physicist Yoshihiko Ōtsuki advanced this plasma theory, adding a scientific dimension to the folklore.

In addition to these scientific explanations, the possibility of optical illusions also cannot be discounted, especially for lights seen moving in the darkness from afar.

The complexity and variety of Onibi legends make it difficult to attribute a single explanation to all observed phenomena. Adding to the confusion is the frequent conflation of Onibi with other similar folkloric entities like hitodama (souls of the dead) and kitsunebi (fox fires). The myriad of theories, both scientific and supernatural, reflects the rich tapestry of folklore and the human endeavor to understand the unexplainable. As such, the true nature of Onibi remains shrouded in mystery, a captivating blend of cultural legend and scientific inquiry.

Tenka

The Tenka, a mesmerizing yet mysterious phenomenon in Japanese folklore, captivates with its enigmatic presence across various regions of Japan. These atmospheric ghost lights, with their rich historical roots, are chronicled in several notable works such as the Ehon Hyaku Monogatari from the Edo period and the essay Kasshi Yawa by Seizan Matsuura. Beyond these literary references, the Tenka are deeply ingrained in local lore, each region offering its unique interpretation and tales.

In the Atsumi District of Aichi Prefecture, a fascinating occurrence takes place. When travelers find their path illuminated as if it were noon, despite the cloak of night, they attribute this to the "tenbi", a local variant of the Tenka. This sudden brightness on a dark road is both awe-inspiring and unnerving.

Further, in the Ibi District of Gifu Prefecture, the Tenka take on a slightly different, more dynamic character. Here, they are known as "tenpi", characterized by mysterious fires that crackle and roar through the evening sky in summer, darting about like celestial dancers.

The Higashimatsuura District in Saga Prefecture has its unique twist to the tale. When a Tenka appears, it is believed to herald better weather. However, this blessing comes with a curse, as homes visited by these lights often suffer from illness. To ward off this misfortune, locals resort to striking a kane, a type of gong, to drive the Tenka away.

In the Tamana District of Kumamoto Prefecture, the Tenka are visualized as celestial fires, comparable in size to a paper lantern, mysteriously descending from the heavens. Legend has it that if these ghostly fires touch the roof of a house, a great conflagration ensues. Similarly, in other parts of Saga Prefecture, the appearance of Tenka is feared as an ominous sign, a premonition of impending fires, and is therefore greatly detested.

The Tenka, shrouded in the mysteries of Japanese folklore, are not just mere atmospheric phenomena; they carry with them tales of deep-seated emotions and retribution. These captivating tales are eloquently captured in the "Amakusatō Minzokushi," a document detailing folk customs from the Amakusa Islands in Kumamoto Prefecture. This document narrates a particularly evocative legend, which intertwines the natural and supernatural, offering a glimpse into the complex tapestry of human beliefs and fears.

According to the legend, a man once journeyed to the village of Oniike with intentions to fish. However, the villagers, wary of outsiders, treated him with disdain and hostility. This unwelcoming attitude, combined with the harsh conditions, eventually led to the man's demise from illness. His death, however, was not the end of his presence in Oniike.

Following this tragic event, every evening, a mysterious ball of fire began to appear, soaring through the skies of Oniike. This fireball, with a mind of its own, would land and ignite a brush, rapidly spreading flames that the villagers could not extinguish. House after house was consumed by the fire, leaving the village in ruin. The villagers, stricken with fear and guilt, interpreted this fiery devastation as the wrath of the deceased man's vengeful spirit, an onryō, seeking retribution for the injustice he faced.

In an attempt to appease the spirit and quell their own fears, the villagers erected a jizō statue at the site where the man was mistreated. They began a tradition of mourning and honoring his spirit every winter, hoping to find peace and forgiveness for their past actions.

Adding another layer to the Tenka lore, there is a theory about the sound these ghostly flames make when they fly. It is believed that the Tenka emit a "shan shan" sound, reminiscent of the janjanbi phenomenon from Nara Prefecture. This similarity has led to the Tenka being referred to as "shanshanbi" in the Tosa Province, now known as Kōchi Prefecture. This auditory aspect of the Tenka adds a haunting auditory dimension to their visual spectacle, further enriching the folklore surrounding these enigmatic ghost lights.

Each element of this legend—from the man's tragic fate to the villagers' response and the eerie sounds of the Tenka—combines to create a narrative that is both haunting and poignant. It reflects the intricate relationship between the natural world and the spiritual beliefs in Japanese culture, where folklore often serves as a mirror to human emotions, fears, and desires.

The "Kasshi Yawa," a notable work on this subject, delves into how the people of Saga Prefecture interacted with these enigmatic fireballs. According to this text, when the locals encountered a Tenka and neglected it, calamity ensued in the form of their homes catching fire. To counteract this, they resorted to a communal ritual of sorts. Gathering around the Tenka, they would recite Buddhist prayers, a spiritual chorus that seemed to influence the Tenka. Remarkably, these collective incantations would cause the Tenka to alter its course, as if repelled by the power of faith, and eventually, it would disappear into the vegetation on the outskirts of the town.

Another intriguing method believed to repel these fiery apparitions involved the use of a setta, a traditional Japanese sandal. This unusual technique finds mention in the "Fude no Susabi," a collection of fantastic stories from the Ansei period. A tale from this collection recounts the plight of a resident from the Hizen Province, who suffered the loss of their home to a fire caused by a Tenka. In a dramatic twist, as the Tenka shifted to another person's home, the quick-thinking resident fended it off with a setta. Unfortunately, this action led to the original victim's house catching fire again. This incident escalated to a dispute where the initial victim sought compensation for their lost home from the daikan, the administrative officials of the Edo period.

The "Ehon Hyaku Monogatari," another Edo period collection of fantastic stories, also delves into the mystique of the Tenka. It recounts how these fireballs were linked to untimely deaths and devastating fires in homes. One particularly gripping story from this collection involves a heartless daikan known for his harsh treatment of subordinates and poor standing with his superiors. His fate was sealed when, shortly after stepping down from his position, a mysterious fire, originating from an inexplicable source, engulfed his home. Tragically, he perished in the blaze, and his wealth—gold, silver, and all his possessions—were reduced to smoke in mere moments. This catastrophic event was marked by an eyewitness account of a fiery lump plummeting from the sky, a detail that adds an ominous and otherworldly aspect to the tale.

These stories, varying in their details and the human interactions with the Tenka, paint a vivid picture of a time when mystical beliefs and practices were deeply intertwined with daily life. They reflect a cultural landscape where the unexplained was often attributed to supernatural forces, and where rituals, prayers, and even simple objects like sandals were imbued with the power to influence these mysterious elements of nature.

Min Min Lights

The lore of the Min Min lights is a tapestry of stories and beliefs, deeply rooted in the rich cultural heritage of Aboriginal Australian societies. These tales, which predate the European colonization of Australia, have woven themselves into the broader fabric of Australian folklore. Indigenous Australians, particularly those who have witnessed these enigmatic phenomena, often speak of an increase in sightings correlating with the European expansion into the remote outback. The first documented sighting of these mysterious lights is often attributed to an account from 1838, detailed in the book "Six Months in South Australia." However, there is a possibility that this specific event might describe a different, yet equally intriguing, natural phenomenon.

The etymology of the term 'Min Min' is shrouded in mystery. One theory suggests a connection to an Aboriginal language from the Cloncurry region, while another links it to the Min Min Hotel, a landmark in a small settlement where a stockman reported witnessing the light in 1918. Despite these theories, a definitive origin of the name remains elusive.

The tales woven around the Min Min lights by non-Indigenous Australians often paint them as a cryptic and supernatural occurrence. These narratives typically align with the broader context of the Australian Gothic, imbuing the landscape with a sense of eerie mystery and otherworldliness. The Min Min lights, in these stories, are often portrayed as benign yet unsettling, leaving those who encounter them with a sense of awe and trepidation.

Reports of these lights span the vast expanse of Australia, from Brewarrina in western New South Wales to Boulia in northern Queensland, with a significant concentration of sightings in the Channel Country. Yunta, South Australia, a region known for its extreme heat and situated in a low-lying basin, is another hotspot for these mysterious sightings. The Ngarluma people of the Pilbara region in Western Australia, particularly around Pyramid Station, also recount experiences with the Min Min lights, adding another layer to this captivating phenomenon that continues to intrigue and mystify.

The phenomenon of the Min Min lights is a kaleidoscope of experiences and descriptions, each account adding its unique hue to this enigmatic Australian mystery. Most commonly, witnesses describe these lights as fuzzy, disc-shaped illuminations that seem to float eerily just above the horizon, as if defying the very laws of physics. The lights exhibit a spectral dance of colors, often starting as a ghostly white before morphing through a spectrum of reds and greens, adding to their otherworldly appearance.

The intensity of these lights is as variable as their color. Some accounts speak of a dim, almost ethereal glow, while others recall an intense brightness, so potent that it casts stark, clearly defined shadows and bathes the ground beneath in an unearthly light. This variability only deepens the mystery surrounding these lights and fuels the endless fascination with them.

Intriguingly, these lights are not mere passive spectacles. Folklore abounds with tales of the Min Min lights displaying almost sentient behaviors. There are chilling accounts of the lights following or advancing towards people, only to vanish abruptly when confronted or shot at. Yet, these lights are capricious, often reappearing moments later, as if to taunt the observer. This game of hide-and-seek with the unknown has given rise to a dire warning in the tales: those who dare to chase after these lights and somehow catch them are fated never to return, swallowed by the mystery they sought to unravel.

Some witnesses have recounted how the light seemed to engage in a ghostly dance, approaching them repeatedly, only to retreat each time, as if luring them into the unknown. Others have reported a more unsettling experience, where the lights managed to keep pace with their moving vehicles, a haunting escort through the remote Australian landscape. These accounts contribute to a rich tapestry of stories and experiences, making the Min Min lights a captivating enigma in Australian folklore.

The enigma of the Min Min lights, a phenomenon that has captured the imagination and curiosity of many, remains shrouded in mystery. The question of whether these lights are a tangible, real-world occurrence and, if so, what their origin might be, continues to puzzle scientists and enthusiasts alike. Several hypotheses have been proposed in an attempt to shed light on this perplexing phenomenon.

One intriguing theory involves the concept of bioluminescence. Scientist Jack Pettigrew has put forth the idea that these lights could be the result of swarming insects that have acquired bioluminescent properties. This transformation, he suggests, might occur due to contamination from naturally occurring agents found in local fungi. Another variant of this theory proposes the involvement of a species of owl equipped with a natural bioluminescent source. Despite these compelling theories, no concrete evidence has been found; no such animal displaying these characteristics has been captured or observed, and the intensity of the known bioluminescent sources seems insufficient to account for the reported brightness of the Min Min lights.

Another hypothesis delves into the realm of geophysical phenomena. It's speculated that the lights might be a manifestation of piezoelectrics or marsh gas. However, this theory faces a significant challenge, as the Min Min lights are often reported in areas lacking the geological conditions that typically give rise to such phenomena.

A third and fascinating hypothesis suggests that the Min Min lights could be a form of Fata Morgana mirage. A Fata Morgana is an extraordinary type of mirage caused by a drastic temperature difference between layers of air. This optical phenomenon can make distant lights or objects, which are actually beyond the horizon, appear to float above it, often appearing distorted. This theory intriguingly aligns with the historical evolution of the Min Min lights' sightings. The earliest reports, including Aboriginal legends, described stationary lights, which could be explained as refractions of distant campfires. In contrast, more recent accounts tell of lights that seem to move, potentially the refracted images of car headlights, distorted and made mobile by the mirage effect. The desert environment where the Min Min lights are observed is known for atmospheric temperature inversions, which could feasibly create conditions ripe for such mirages.

Each of these theories, while offering a glimpse into potential explanations, also highlights the complexity and enduring mystery of the Min Min lights. They remain an elusive puzzle, a tantalizing blend of folklore, natural wonder, and scientific intrigue, continuing to fascinate and perplex those who seek to unravel their secrets.

Hessdalen Lights

The mysterious Hessdalen lights, a phenomenon shrouded in enigma, have captured the imagination and curiosity of both locals and scientists alike. Originating from an unknown source, these lights manifest in the skies of the Hessdalen valley, both during daylight and under the cover of night. Their appearance is captivating, often emerging as luminescent orbs in shades of bright white, vivid yellow, or deep red. These ethereal lights are not bound by a single behavior; they exhibit a variety of movements and durations. Sometimes, they appear for fleeting moments, lasting just a few seconds, while at other times, they enchant onlookers for over an hour.

The lights display an extraordinary range of motion. On some occasions, they dart across the sky with incredible speed, cutting through the air like swift meteorites. In stark contrast, there are moments when they gently sway back and forth, almost as if dancing to an unseen rhythm. In other instances, they remain suspended in mid-air, defying gravity and expectations alike.

Reports of these unusual lights date back to at least the 1930s, hinting at a long-standing mystery in the region. Between December 1981 and mid-1984, there was a significant surge in activity. During this period, the lights were observed up to 20 times per week, turning the valley into a hub for intrigued tourists and overnight visitors. However, by 2010, the frequency of these sightings had reduced considerably, with only 10 to 20 occurrences reported each year.

In an attempt to unravel this mystery, "Project Hessdalen" was launched in 1983 by UFO-Norge and UFO-Sverige. This initiative marked the beginning of a series of field investigations that continued until 1985. The project brought together a diverse team of students, engineers, and journalists, collectively known as "The Triangle Project." In 1997-1998, this team made a significant observation, recording the lights forming a pyramid shape that bounced rhythmically up and down.

To further these investigations, the Hessdalen Automatic Measurement Station (Hessdalen AMS) was established in 1998. Positioned in the heart of the valley, this station is dedicated to continuously monitoring and recording the appearance of these mysterious lights. Later, a collaborative programme named EMBLA was initiated, aiming to bridge the gap between established scientists and students in researching these enigmatic lights. This program has garnered the involvement of prestigious institutions such as Østfold University College in Norway and the Italian National Research Council, both of which are at the forefront of this intriguing scientific endeavor.

Various theories have been proposed to shed light on this enigmatic occurrence, each offering a potentially plausible explanation yet none gaining universal acceptance.

Some of the sightings of the Hessdalen lights have been rationalized as misinterpretations of more mundane phenomena. Instances have been identified where the lights were actually misperceptions of known and ordinary objects or occurrences, such as astronomical bodies, aircraft, car headlights, and even atmospheric mirages. These explanations, while accounting for a portion of the sightings, do not encompass the entirety of the experiences reported by witnesses.

One particularly intriguing theory posits that the phenomenon is a result of a unique and not fully understood process of combustion. This process involves airborne dust particles, which are abundant in the area due to mining activities. A detailed analysis of these particles revealed the presence of hydrogen, oxygen, and a range of other elements, including titanium. A significant focus of this theory is on the large deposits of scandium found in the Hessdalen region, which are thought to contribute to this combustion process. The publication of this research was met with significant media attention in Norway, leading to headlines proclaiming that the mystery of the Hessdalen lights had been solved.

Another hypothesis, proposed in 2010, offers a more complex scientific explanation. It suggests that the lights are the result of macroscopic Coulomb crystals forming within a plasma. This plasma is theorized to be produced by the ionization of air and dust particles, triggered by alpha particles emitted during the decay of radon in the atmosphere. This theory is supported by several observations related to the Hessdalen lights, such as their oscillation patterns, geometric structures, and specific light spectra. These characteristics could potentially be explained by a dust plasma model. The decay of radon, a process known to produce alpha particles and radioactive elements like polonium, is a crucial aspect of this hypothesis. In 2004, physicist Massimo Teodorani conducted research that aligned with this theory, detecting higher levels of radioactivity on rocks near locations where large light balls had been reported. Furthermore, computer simulations have shown that dust particles, when immersed in ionized gas, can organize themselves into structures like double helixes, a formation occasionally observed in the Hessdalen lights.

Yet another hypothesis delves into the realm of piezoelectricity. It suggests that the Hessdalen lights could be a byproduct of piezoelectric effects induced under specific rock strains. This theory gains credibility due to the presence of quartz-rich rocks in the Hessdalen valley. Quartz, known for its piezoelectric properties, could produce an intense charge density under certain conditions, potentially leading to the creation of the mysterious lights.

In their 2011 research paper, Gerson Paiva and Carlton Taft delved deeper into the mysteries of the Hessdalen lights, particularly focusing on the dusty plasma theory. Their study provided a fresh perspective on the phenomenon, especially in relation to the role of piezoelectricity in quartz, which was previously thought to be a contributing factor. They argued that this theory of piezoelectricity could not adequately explain one of the most peculiar and intriguing aspects of the Hessdalen lights – the formation of geometrical structures at the center of the phenomenon.

Paiva and Taft's research brought forward a new mechanism to explain the formation of light ball clusters, a hallmark of the Hessdalen lights. They proposed that these clusters are formed through the nonlinear interaction of ion-acoustic and dusty-acoustic waves with low-frequency geoelectromagnetic waves within dusty plasmas. This interaction, they theorized, leads to the creation of the light balls that are characteristic of the Hessdalen phenomenon. The velocity of these ejected light balls was a key focus of their study. Theoretically, they calculated this velocity to be around 10,000 meters per second, which intriguingly aligns well with the observed velocity of some ejected light balls, estimated at a staggering 20,000 meters per second.

A distinctive feature of the Hessdalen lights, as observed in their manifestations, is the color variation between the central ball and the ejected balls. The central ball typically appears white, while the ejected balls display a consistent green color. Paiva and Taft ascribed this color differentiation to the radiation pressure produced by the interaction between very low-frequency electromagnetic waves and atmospheric ions. This interaction occurs predominantly in the central white-colored ball through ion-acoustic waves. They proposed that $O+2$ ions, specifically undergoing an electronic transition that results in green emission lines, are likely the only ions transported by these waves. This concept is supported by the presence of electronic bands of $O+2$ ions in auroral spectra, providing a potential link between the observed green light balls and known atmospheric phenomena.

This comprehensive study by Paiva and Taft not only challenges previous theories but also adds a significant layer of complexity and understanding to the enigmatic Hessdalen lights phenomenon. Their work underscores the multifaceted nature of this mystery, suggesting that a combination of plasma physics, atmospheric chemistry, and electromagnetic interactions might be at play in creating the spectacular light displays observed in the Hessdalen valley.

One of the critical factors under investigation is the estimated temperature of these luminous manifestations, which hovers at an astonishing 5,000 Kelvin (equivalent to 4,730 degrees Celsius or 8,540 degrees Fahrenheit). This scorching temperature is a key element in understanding the underlying processes at play in the formation of the Hessdalen lights, particularly in relation to the behavior of ionized particles.

At this extreme temperature, the rate coefficients of dissociative recombination come into play. Specifically, the oxygen ions exhibit a rate coefficient of 10^{-8} cm^3 s^{-1}, while the nitrogen ions display a rate coefficient of 10^{-7} cm^3 s^{-1}. What makes this information significant is that it highlights a critical distinction: in the plasma of the Hessdalen lights, nitrogen ions undergo decompression at a significantly faster rate than oxygen ions. It's worth noting that only ionic species are effectively transported by ion acoustic waves in this context. Consequently, this disparity in recombination rates results in the dominance of oxygen ions in the ejected green light balls observed in the Hessdalen lights. These oxygen ions exhibit a unique electronic transition that produces the characteristic negative band of O^+_2, contributing to the green glow witnessed in these mysterious phenomena after the formation of ion-acoustic waves.

Paiva and Taft furthered their investigation by presenting a model aimed at resolving the seemingly contradictory spectral characteristics observed in the Hessdalen lights. The spectrum, as described, is characterized by a nearly flat top with steep sides, a feature attributed to the influence of optical thickness on the bremsstrahlung spectrum. At lower frequencies, self-absorption comes into play, causing the spectrum to follow the Rayleigh–Jeans portion of the blackbody curve. This peculiar spectral profile is a hallmark of dense ionized gas.

Furthermore, the spectrum generated by the thermal bremsstrahlung process remains flat up to a cutoff frequency before experiencing an exponential decline at higher frequencies. This sequence of events is instrumental in forming the distinctive spectrum associated with the Hessdalen lights phenomenon, particularly when the atmosphere is clear and devoid of fog.

To add another layer of complexity to the phenomenon, Paiva and Taft's model suggests that the spatial color distribution of the luminous balls frequently observed in the Hessdalen lights is a result of electrons being accelerated by electric fields during the rapid fracture of piezoelectric rocks beneath the ground. This concept introduces a fascinating interplay between geological elements and the generation of the luminous phenomena. In 2014, Jader Monari contributed to this body of research by presenting a new Hessdalen lights model involving a geological-like battery scenario. In this intriguing model, the two sides of the valley serve as electrodes, while the river Hesja functions as the electrolyte. Gas bubbles rising into the air can become electrically charged, ultimately producing gas luminescence and contributing to the captivating and mystifying Hessdalen lights phenomenon.

In summary, the research into the Hessdalen lights takes us on a journey through the complexities of ionized plasmas, optical physics, and the geological dynamics of the region, all of which converge to create the mesmerizing display of lights witnessed in the Hessdalen valley. Each layer of investigation adds depth to our understanding of this phenomenon, yet it remains a captivating mystery, beckoning researchers to explore its intricacies further.

Martebo Lights

The enigmatic Martebo Light, or Marteboljuset, is a mysterious atmospheric phenomenon often observed in the Martebo boglands of Gotland, Sweden. This 'ghost light' has been a subject of fascination and speculation for over a century, with its first recorded sighting dating back to March 1922. The light is often described as a large, floating orb several meters above the ground, with varying accounts of its appearance. Some eyewitnesses report it flickering like a flame, while others describe it as a steady, glowing ball of light.

What sets the Martebo Light apart from other anomalous lights is its associated legend, which adds a haunting and poignant narrative to the sightings. The tale revolves around a man named Knut Stare, who resided in Knutstorp with his son. The story goes that one night, a group of soldiers sought shelter and sustenance at Stare's home. After an evening of camaraderie and drinking, Stare fell into a deep slumber, only to awaken to the disappearance of both his son and the soldiers. The legend suggests that the Martebo Light symbolizes the restless spirit of Knut Stare, eternally searching the boglands for his lost son.

However, not everyone is convinced of the supernatural nature of the Martebo Light. Skeptics and scientists have offered more grounded explanations, suggesting that the lights might be the result of mundane sources like flashlights or car headlights from nearby roads. This theory has been bolstered by experiments conducted by UFO-Sweden, where test subjects mistook deliberately placed car headlights for the ghostly light. Other rational explanations include natural phenomena such as ball lightning or the emissions of swamp gas, which is plausible given Martebo's boggy terrain. Despite these scientific interpretations, the mystery and allure of the Martebo Light continue to captivate locals and visitors alike, keeping the legend alive in the heart of Gotland.

On the evening of November 9, 1993, Ingemar Flodin recounts a remarkable event that unfolded and would leave a lasting impression on those who experienced it. At 6 p.m., Barbro Leion, who would later become Flodin's partner, was approached by their neighbor, Britta Jacobsson. Britta had noticed something unusual at Martebomyr, a nearby bog, and was keen to investigate further. She invited Barbro and Flodin to join her in observing this peculiar phenomenon. Initially, Flodin was skeptical, having heard similar stories about the bog before and not placing much faith in them.

However, Barbro returned about an hour and a half later, visibly moved and excited by what she had seen. She described witnessing a strange and unexplainable light phenomenon at the bog. Her vivid and earnest account persuaded Flodin to reconsider his skepticism. Subsequently, Flodin agreed to accompany Barbro back to Martebomyr in their car to personally witness the phenomenon.

They journeyed down a narrow, winding forest road that led to the bog, eventually parking their car about 300 meters from the main road to Stenkyrka, with Stenkyrkavägen behind them at their observation point. Despite the darkness of the narrow road, the sky appeared somewhat brighter against the dark forest on both sides.

During their observation, an intensely bright light suddenly appeared from the forest to the left of the driver's seat, at an estimated distance of 100 meters. This light shone brilliantly for about four to five seconds before extinguishing. Flodin and Barbro observed these light phenomena for approximately an hour and a half, deeply intrigued by the mysterious lights.

After thoroughly observing the lights, they turned their car around on a nearby exit road and returned to their house in Tingstäde. The experience at Martebomyr that evening was both mystifying and unforgettable, leaving a lasting impression on Flodin and sparking a deeper curiosity about the unexplained mysteries of the world.

Paasselkä Devils

In an era where logical and scientific explanations are sought for almost every phenomenon, the mysterious light occurrences at Paasselkä Lake in Finland captivate the imagination of many. These lights, often reported and witnessed, have no definitive scientific explanation yet, fueling a sense of wonder and speculation about spirits and ghosts. This adds an intriguing, almost supernatural element to the lake's already fascinating history.

Paasselkä Lake, known for its distinctive geological characteristics, is believed to have been formed by a meteorite impact. This hypothesis was long held by specialists, and it was finally substantiated by geological studies conducted in 1999. These studies revealed that this vast depression, now filled with water, was indeed the result of a colossal meteorite striking the Earth. This event occurred approximately 229 million years ago, during the middle–late Triassic period, marking a significant moment in the planet's geological history.

The lake itself is striking in its appearance. It boasts a roughly circular shape, a unique feature among natural lakes. One of its most notable characteristics is the absence of islands in its central part, which is quite uncommon. Adding to its allure is its depth, plunging to about 75 meters at its deepest point, making it one of the deeper lakes in the region. Furthermore, Paasselkä Lake is not isolated; it is connected through several channels to other lakes in the vicinity, being a part of the expansive and picturesque Greater Saimaa lake complex. This interconnectedness contributes to the lake's ecological significance and its role in the local landscape.

Together, the lake's mysterious light phenomena and its significant geological history create a fascinating narrative that blends the mysteries of the natural world with the wonders of ancient cosmic events. This unique combination makes Paasselkä Lake a place of intrigue and scientific interest, attracting researchers, nature enthusiasts, and those curious about the unexplained mysteries of our world.

The enchanting light phenomena over Paasselkä Lake, a subject of awe and mystery, have been chronicled in historical accounts dating back to the 18th century. However, it was the publication of a captivating book in 2006 by Sulo Strömberg, a local history enthusiast, that reignited widespread interest in these mysterious lights. Strömberg's book is a treasure trove of narratives, compiling eyewitness accounts and personal testimonies of those who have witnessed the mesmerizing lights, often referred to as the "Paasselkä devil."

The descriptions of the Paasselkä devil, as recounted by those who have seen it, are both fascinating and varied. Most commonly, the phenomenon appears as a glowing orb, radiating with an intensity that captivates and mystifies. Witnesses describe it as a ball of bright light, often likened to the size of a football. However, estimating its exact dimensions is challenging, as it is usually observed from a considerable distance, silently hovering over the lake's surface.

The colors of these luminous orbs add another layer to their allure. Predominantly, they manifest as white lights, but there have been instances where red orbs have been spotted, adding a hint of otherworldly mystique to the phenomenon. The intensity of the light is striking, with many eyewitnesses noting that it far surpasses the brightness of distant car headlamps that sometimes illuminate the lake's surroundings.

Interestingly, while these lights are most frequently observed over the lake's waters, they have also been sighted in the surrounding areas, including nearby forests and marshes. This suggests a mobility or range to the phenomenon that extends beyond the lake itself. Additionally, while typically seen as solitary orbs, there have been occasions where multiple lights appear simultaneously, creating a spectacular and somewhat eerie tableau against the night sky.

Locals recount tales of the lights displaying seemingly intentional actions. There are reports of these lights following the boats of fishermen, creating an eerie atmosphere on the lake. In some instances, the lights appear to react to human presence, such as moving away when caught in the beam of a torch. This unpredictable behavior lends a mystical quality to the phenomenon, further deepening the intrigue surrounding it.

One notable account from the early 20th century vividly illustrates the mysterious nature of these lights. A group returning by boat from a wedding spotted one of these enigmatic lights hovering above the lake. Driven by curiosity, they attempted to approach it, only to witness the light dart away several hundred meters in a matter of moments, then resuming its calm demeanor as if nothing had happened.

Historically, the local population regarded these lights with a mix of awe and apprehension, often considering them to be malevolent spirits. This belief led to various rituals and offerings made to appease the Paasselkä devils, a practice that persisted into the early 20th century. These historical accounts and traditions add a rich cultural layer to the phenomenon, embedding it deeply in the local folklore and traditions.

In contemporary times, the allure of the Paasselkä devil has been harnessed by tourist companies, adding an intriguing element to their usual fishing tours. Visitors are often thrilled by the prospect of witnessing this elusive and mysterious phenomenon for themselves. Sightings by tourist groups are not uncommon, with some even capturing images and videos of these enigmatic lights, further fueling interest and speculation.

Despite numerous theories and scientific investigations, a convincing explanation for the Paasselkä devil remains elusive. Hypotheses have ranged from the release of gases from lake sediments to rare piezoelectric effects, but none have conclusively unraveled the mystery. This unexplained phenomenon is not unique to Paasselkä Lake; similar light occurrences are reported in various parts of the world, such as the Chir Batti in Gujarat, India, adding a global dimension to the intrigue of these mysterious lights. The continued fascination with these phenomena reflects humanity's enduring interest in the mysteries of the natural world, where some wonders remain beyond the reach of current scientific understanding.

Ball Lightning

B all lightning, a mysterious and elusive natural phenomenon, has captivated observers and scientists alike for centuries. Often seen during thunderstorms, these glowing orbs range in size from tiny, pea-sized specks to imposing structures several meters across. Unlike the fleeting flash of typical lightning, these luminous spheres can persist, lingering far longer than the split-second spark of their more common electrical counterparts. They also differ fundamentally from St. Elmo's fire, another weather-related luminescence.

Historical accounts, dating back to the 19th century and even earlier, are rich with descriptions of these enigmatic balls of light. Intriguingly, some of these tales recount how the balls eventually explode, leaving a distinct sulfuric scent in their wake. Over the years, the phenomenon of ball lightning has appeared in numerous accounts, drawing the keen interest of the scientific community.

In a groundbreaking development, the optical spectrum of an event believed to be ball lightning was captured and published in January 2014. This groundbreaking study not only described the phenomenon but also included high-frame-rate video footage, offering unprecedented insights into its nature.

Laboratory experiments have managed to create visual effects similar to ball lightning, adding another layer of intrigue to its study. These experiments have produced fascinating results, though the connection between these artificial recreations and the actual phenomenon observed in nature remains a topic of debate and research.

Numerous hypotheses have been proposed by scientists over the years to explain the sightings and reports of ball lightning. Despite these efforts, concrete scientific data on the phenomenon is scarce, and its existence as a distinct physical entity remains a subject of speculation and ongoing investigation. The primary basis for the presumption of its existence stems from public sightings, which, while numerous, often yield inconsistent and non-reproducible findings. Consequently, the reality of ball lightning, as a unique and separate natural occurrence, continues to be an unsolved mystery, intriguing experts and laypeople alike with its elusive and captivating presence.

Historical Attestations

Ball lightning has long been a source of fascination and mystery, inspiring various legends and myths worldwide, including the mythological Anchimayen from the Mapuche culture of Argentina and Chile.

Statistical studies have highlighted the widespread nature of ball lightning sightings. In 1960, it was estimated that around 5% of the global population had witnessed this phenomenon. Another extensive study delved into around 10,000 reported cases of ball lightning, underscoring its prevalence and the intrigue it generates.

One of the earliest potential references to ball lightning can be found in the chronicles of Gervase of Canterbury, an English monk, dated 7 June 1195. Gervase described a remarkable event near London involving a dense, dark cloud that emitted a white substance. This substance then formed into a spherical shape beneath the cloud, eventually transforming into a fiery globe that descended towards a river.

This account caught the attention of Emeritus Professor Brian Tanner, a physicist, and historian Giles Gasper from Durham University. They suggested that Gervase's description might be one of the earliest known references to ball lightning. The narrative of a white substance emerging from a cloud, forming into a spinning, fiery sphere, and exhibiting some horizontal movement closely aligns with both historical and modern reports of ball lightning. The researchers found it intriguing how Gervase's 12th-century account mirrored contemporary descriptions of this enigmatic phenomenon.

One notable incident occurred during the Great Thunderstorm at Widecombe-in-the-Moor in Devon, England, on 21 October 1638. A severe storm led to a tragic event where four people died, and about 60 were injured. According to eyewitnesses, an enormous ball of fire, about 8 feet in diameter, struck the church, causing near-catastrophic damage. It hurled large stones from the church walls to the ground, shattered pews and windows, and filled the church with a pungent sulphurous smell and thick smoke.

This ball of fire reportedly split into two parts; one smashed through a window to escape, and the other vanished inside the church. The incident, marked by fire and sulphur, led contemporaries to attribute it to supernatural causes like "the devil" or "flames of hell". Some even blamed the tragedy on two individuals who were playing cards in the church during a sermon, supposedly invoking divine wrath.

In another instance, in December 1726, John Howell of the sloop Catherine and Mary recounted a striking event in a letter. As they navigated through the Gulf of Florida on 29th August, a large ball of fire descended from the sky, causing extensive damage to the ship. It splintered the mast into countless pieces, broke the main beam and several planks, and resulted in one fatality and severe injuries to another crew member. The incident was so intense that, if not for heavy rains, their sails might have caught fire.

A remarkable report from 1749 involves HMS Montague. Admiral Chambers, aboard the ship, witnessed a large ball of blue fire. Despite efforts to avoid it, the fireball exploded near the ship, shattering the main top-mast and causing the main mast to collapse. The explosion, likened to the discharge of a hundred cannons, left a strong sulphurous odor. Five men were knocked down, with one severely injured.

A deadly encounter with ball lightning was recorded in 1753 involving Professor Georg Richmann in Saint Petersburg, Russia. Inspired by Benjamin Franklin's experiments, Richmann was conducting a kite experiment during a storm. A ball of lightning traveled down the kite's string, striking Richmann's forehead and killing him instantly. The lightning left a red mark on his forehead, blew open his shoes, singed his clothes, and caused extensive damage to his room, leaving his engraver unconscious.

During a storm in 1809, the British ship HMS Warren Hastings had a dramatic encounter with ball lightning, as reported in an English journal. The ship's crew witnessed three separate "balls of fire." The first descended onto the deck, tragically killing a man and setting the main mast ablaze. In a shocking turn of events, a crewman attempting to recover the body was struck by a second ball, which threw him back and left him with mild burns. A third ball lightning strike resulted in the death of another crew member. After these incidents, the crew noted a lingering, unpleasant smell of sulphur.

Ebenezer Cobham Brewer, in the 1864 US edition of his book "A Guide to the Scientific Knowledge of Things Familiar," discusses "globular lightning." Brewer describes these as slow-moving fireballs or explosive gases that may occur during thunderstorms, sometimes falling to the ground or skimming across it. He noted that these balls could split into smaller ones and potentially explode with force comparable to a cannon.

Wilfrid de Fonvielle, a French science writer, provided further insights into this phenomenon in his book "Thunder and Lightning," translated into English in 1875. He mentioned around 150 reports of globular lightning, noting its apparent attraction to metals and its ability to appear in various colors. For instance, in Coethen in the Duchy of Anhalt, it was reported as green. He recounted several intriguing incidents, such as a ball of lightning descending along a poplar tree's bark, bouncing upon touching the ground, then vanishing without explosion. Another report described a ball of lightning rolling harmlessly through a kitchen in Salagnac, only to explode in an adjacent stable, fatally injuring a pig. De Fonvielle highlighted that although these balls of lightning might move slowly and even pause, their potential for destruction was significant. He cited an example where a ball of lightning that entered the Stralsund church exploded, projecting several smaller balls that also exploded like shells.

In a captivating account, Tsar Nicholas II, the last emperor of Russia, recounts a spellbinding experience he had as a young boy with his grandfather, Emperor Alexander II. During an all-night vigil in a small church, a fierce thunderstorm erupted. Amidst the flashes of lightning and thunderous roars, a dark, ominous atmosphere enveloped the church. Suddenly, a fiery orb, resembling ball lightning, burst through the window, swirling towards Emperor Alexander II. This electrifying sphere danced across the floor, soared around the chandelier, and then whisked out the door into the park. Despite the initial shock, Nicholas found reassurance in his grandfather's calm demeanor, who faced this awe-inspiring phenomenon with a serene acceptance, instilling a sense of courage and faith in the young Tsar.

British occultist Aleister Crowley also shared a remarkable experience during a thunderstorm on Lake Pasquaney in New Hampshire in 1916. While sheltered in a cottage, he was astounded to see a dazzling globe of electric fire, about six to twelve inches in diameter, hovering near him. This sphere of what he called "globular electricity" exploded with a loud bang, distinctly different from the storm's chaos outside. Crowley experienced a mild shock, adding a tactile dimension to this extraordinary encounter.

Adding to these intriguing accounts, R.C. Jennison from the University of

Kent's Electronics Laboratory detailed his observation of ball lightning in a 1969 article in Nature. He was a passenger on an Eastern Airlines flight from New York to Washington when the aircraft flew into an electrical storm. Following a bright and loud electrical discharge, Jennison witnessed a glowing sphere, over 20 cm in diameter, glide down the airplane's aisle, passing close to him. This enigmatic orb maintained a consistent height and trajectory throughout its visible journey.

Willy Ley recounted an extraordinary event from Paris on July 5, 1852, which was officially recorded with the French Academy of Science. During a thunderstorm, a tailor living near the Church of the Val-de-Grâce witnessed a ball lightning the size of a human head emerge from his fireplace. This mysterious orb zipped around the room before returning to the fireplace, where it exploded and destroyed the chimney's top.

Another notable incident took place on April 30, 1877, at the revered Golden Temple in Amritsar, India. A ball of lightning entered and exited through a side door, leaving several witnesses in awe. This event was significant enough to be inscribed on the front wall of the Darshani Deori.

In Golden, Colorado, on November 22, 1894, an unusual and extended display of natural ball lightning occurred, hinting at the possibility of artificially inducing such phenomena. The Golden Globe newspaper vividly described balls of fire playing around the School of Mines' new Hall of Engineering. This remarkable occurrence, witnessed by many, involved balls of fire frolicking in the wind-filled, electrically charged air.

On May 22, 1901, in the Kazakh city of Ouralsk (now Oral, Kazakhstan), a "dazzlingly brilliant ball of fire" descended slowly from the sky amid a thunderstorm. It entered a house where 21 people had sought shelter, causing chaos and destruction. The ball lightning broke through a wall, shattered a stove-pipe, and exited through a broken window. This incident was reported the following year in the Bulletin de la Société astronomique de France.

In a dramatic episode in July 1907, ball lightning struck the Cape Naturaliste Lighthouse in Western Australia. The lighthouse keeper, Patrick Baird, was inside the tower at the time and was rendered unconscious. His daughter Ethel documented this frightening experience, adding to the myriad of mysterious ball lightning accounts.

Willy Ley recounted a fascinating event in Bischofswerda, Germany, on April 29, 1925. Witnesses observed a silent ball of light land near a mailman, travel along a telephone wire to a school, and dramatically push back a teacher who was using the telephone. This ball lightning also created perfectly round, coin-sized holes in a glass pane. The incident resulted in 210 meters of melted wire, damaged telephone poles, a broken underground cable, and several workmen being thrown to the ground, though miraculously unharmed.

Laura Ingalls Wilder, in her historically inspired children's books set in the 19th century, includes an early fictional reference to ball lightning. Wilder insisted these stories were based on her real-life experiences. In one scene, during a winter blizzard, three balls of lightning emerge near the family's stovepipe, roll across the floor, and then vanish as Caroline Ingalls, the mother, attempts to chase them with a broom.

During World War II, pilots reported an unusual phenomenon resembling ball lightning, known as foo fighters. These small, luminous balls moved in peculiar trajectories, bewildering the pilots who witnessed them.

Submariners in the same war frequently reported ball lightning in the confined spaces of their submarines. These incidents often occurred when battery banks were switched on or off, particularly if done incorrectly, or when electrical motors were improperly connected or disconnected. Attempts to recreate these phenomena later resulted in several failures and an explosion.

On August 6, 1994, in Uppsala, Sweden, ball lightning is believed to have passed through a closed window, leaving a circular hole about 5 cm in diameter.

Discovered days later, the incident was thought to have occurred during a thunderstorm, supported by local residents' accounts and data from Uppsala University's lightning research division.

In 2005, Guernsey witnessed an incident where an aircraft apparently struck by lightning led to multiple sightings of fireballs on the ground.

A dramatic episode occurred on July 10, 2011, in Liberec, Czech Republic, during a severe thunderstorm. A ball of light with a two-meter tail entered a window into the control room of local emergency services. The ball bounced around the room, rolling along the floor before vanishing. The staff smelled burning and found their communication equipment knocked out, with only a computer monitor destroyed.

On December 15, 2014, Loganair Flight 6780 in Scotland experienced ball lightning in the forward cabin moments before the aircraft was struck by lightning. The plane plunged several thousand feet and narrowly avoided crashing into the North Sea, eventually making an emergency landing at Aberdeen Airport.

Finally, on June 24, 2022, during a massive thunderstorm in Liebenberg, Lower Austria, a retired lady witnessed a blinding lightning strike followed by a yellowish, flame-like object moving along a local road. This ball lightning, recorded by the European Severe Storms Laboratory, displayed a wavy trajectory and disappeared after 2 seconds.

Studies and Experiments

Ball lightning, exhibits a wide array of behaviors and characteristics that have fascinated observers. Its movement is unpredictable – it can glide smoothly or erratically, ascend, descend, or even remain stationary, interacting with the environment in perplexing ways. It might follow the wind or move against it, and its relationship with objects like buildings, vehicles, and people ranges from attraction to indifference or repulsion. Intriguingly, some accounts suggest it can pass through solid materials like wood or metal without impact, while others report destructive capabilities, causing melting or burning.

Often confused with St. Elmo's fire, ball lightning is a distinct phenomenon. St. Elmo's fire is a continuous electrical spark, while ball lightning is a transient event, each unique in their occurrence and properties.

The way ball lightning vanishes is as varied as its appearance. It might simply fade away, burst suddenly, or even explode violently, with some reports suggesting damage. The danger it poses to humans is equally uncertain, with accounts ranging from entirely harmless to potentially lethal.

A comprehensive review in 1972 synthesized these varied accounts into a profile of "typical" ball lightning, while cautioning against the reliance solely on eyewitness testimonies. This profile describes ball lightning as often spherical or pear-shaped with indistinct edges, ranging in size from a few centimeters to a meter. The brightness of ball lightning is comparable to a domestic lamp, making it visible even in daylight. It displays a spectrum of colors, predominantly red, orange, and yellow, and maintains a fairly consistent brightness throughout its lifespan, which can vary from a fleeting second to over a minute.

Ball lightning's movement is generally horizontal at a few meters per second, but it can also move vertically, hover, or meander unpredictably. Some reports describe a rotational motion. It's rare for observers to feel heat from ball lightning, although its disappearance can sometimes release heat.

An interesting aspect of ball lightning is its interaction with metal objects and conductors, often following along wires or fences. It has been observed inside buildings, passing through barriers like doors and windows. In some cases, ball lightning has appeared inside metal aircraft, entering and exiting without causing damage.

The end of a ball lightning event is usually abrupt, occurring either silently or with an explosion. Witnesses often report odors resembling ozone, burning sulfur, or nitrogen oxides following these incidents.

In a groundbreaking study, scientists from Northwest Normal University in Lanzhou, China, captured a rare phenomenon in July 2012 on the Tibetan Plateau – the optical spectrum of what they believed to be natural ball lightning. This significant observation occurred during a study of conventional cloud-to-ground lightning. The team managed to record 1.64 seconds of digital video, capturing the ball lightning from its formation post a regular lightning strike to its gradual fade. An additional high-speed video, shooting at 3000 frames per second, recorded the final 0.78 seconds of the event.

Equipped with slitless spectrographs, both cameras documented this rare occurrence. The analysis revealed emission lines of neutral atomic elements like silicon, calcium, iron, nitrogen, and oxygen. This finding contrasts with the spectrum of the parent lightning, which predominantly showed ionized nitrogen emission lines. The ball lightning moved horizontally at an average speed of 8.6 meters per second, had a diameter of 5 meters, and traveled approximately 15 meters in the recorded timeframe.

Interestingly, the researchers observed oscillations in the light intensity and emissions from oxygen and nitrogen at a frequency of 100 hertz. This fluctuation might have been influenced by the electromagnetic field of a nearby 50 Hz high-voltage power transmission line. The temperature of the ball lightning was estimated to be lower than that of the parent lightning, at less than 15,000 to 30,000 Kelvin. These observations support theories that ball lightning could involve vaporized soil and that it is sensitive to electric fields.

Notably, Nikola Tesla, the renowned inventor, reportedly had the capability to create artificial ball lightning, producing spheres about 1.5 inches in diameter during demonstrations. However, Tesla's primary focus was on higher voltages, power, and the remote transmission of energy, with the ball lightning phenomena being more of a side curiosity.

The International Committee on Ball Lightning (ICBL), an assembly dedicated to this phenomenon, used to organize symposia discussing these mysterious orbs. Another group, known as "Unconventional Plasmas," also explores similar topics. Unfortunately, the last planned ICBL symposium in San Marcos, Texas, in July 2012, was canceled due to insufficient abstract submissions.

In laboratory experiments, varied approaches have been tried:

- Wave-Guided Microwaves: Researchers Ohtsuki and Ofuruton achieved the creation of "plasma fireballs" using microwave interference within a cylindrical cavity, powered by a high-frequency microwave oscillator.
- High-Voltage Experiments: Some groups, including the Max Planck Institute, reportedly produced ball lightning-like effects by discharging high-voltage capacitors in water tanks.
- Home Microwave Oven Experiments: A popular method involves using microwave ovens to generate small, glowing plasma balls. This is typically done by microwaving a lit match or similar object, leading to the formation of fireballs and plasma balls near the oven's ceiling. In some cases, these experiments use a glass jar to contain the phenomena, though this can lead to the jar's explosion.
- Silicon Experiments: A 2007 experiment involved electrifying silicon wafers, vaporizing the silicon and causing oxidation in the vapors. This resulted in the appearance of small, glowing orbs. Brazilian scientists Antonio Pavão and Gerson Paiva reportedly succeeded in consistently creating small, long-lasting balls using this method, supporting the hypothesis that ball lightning might be related to oxidized silicon vapors.

Ball lightning remains one of nature's most enigmatic phenomena, lacking a widely accepted scientific explanation despite being studied since the mid-19th century by notable figures like William Snow Harris and François Arago. Over the years, various hypotheses have emerged, each attempting to unravel the mystery:

Vaporized Silicon Hypothesis: This theory posits that ball lightning is composed of vaporized silicon, resulting from lightning strikes that vaporize the silica in the soil. This process could separate oxygen from silicon dioxide, creating pure silicon vapor. As it cools, the silicon might condense into an aerosol, glowing as it recombines with oxygen. Experiments conducted in 2007 support this hypothesis, successfully creating luminous balls by evaporating silicon with an electric arc. The 2014 recording of natural ball lightning's spectra provided further evidence, indicating a composition similar to that of soil.

Electrically Charged Solid-Core Model: In this model, ball lightning is thought to have a solid, positively charged core surrounded by a thin layer of electrons. Between the core and this electron layer, an intense electromagnetic field is trapped, preventing the electrons from collapsing into the core.

Microwave Cavity Hypothesis: Proposed by Pyotr Kapitsa, this hypothesis suggests that ball lightning is a type of glow discharge, maintained by microwave radiation that travels along ionized air from lightning clouds. The ball itself acts as a resonant microwave cavity, adjusting its size to maintain resonance with the microwaves.

Handel Maser-Soliton Theory: According to this theory, ball lightning is the result of a massive atmospheric maser. The lightning ball appears as a plasma formation at the antinodal plane of microwave radiation from this maser.

Zhejiang University Research (2017): Researchers from Zhejiang University proposed that ball lightning's glow results from microwaves trapped inside a plasma bubble. This bubble forms at the tip of a lightning stroke, creating a spherical plasma structure that holds the radiation. This theory suggests that the ball's brightness is maintained by ongoing plasma generation and eventually fades as the trapped radiation decays, with potential for explosive destabilization. This hypothesis might explain why ball lightning can occur indoors, as microwaves can pass through glass.

Soliton Hypothesis: This theory, proposed by Julio Rubinstein, David Finkelstein, and James R. Powell, suggests that ball lightning is a form of detached St. Elmo's fire. The hypothesis posits that a free ball of ionized air can amplify the ambient electric field through its conductivity, sustaining its ionization and thus forming a soliton in atmospheric electricity. The size of the ball lightning, according to Powell's kinetic theory, is determined by certain electrical properties of the air near the point of breakdown. This model was further developed to suggest that ball lightning involves spherical oscillations of charged particles within plasma, similar to spatial Langmuir solitons, potentially leading to a superconducting phase inside the ball.

Hydrodynamic Vortex Ring Antisymmetry: Another theory considers the possibility of combustion within a spherical vortex, potentially explaining the diverse observational evidence of ball lightning. This concept involves the breakdown of a natural vortex, like Hill's spherical vortex, and how it might produce ball lightning phenomena.

Nanobattery Hypothesis: Proposed by Oleg Meshcheryakov, this theory envisions ball lightning as a collection of composite nano or submicrometer particles, each acting as a tiny battery. A surface discharge could short these batteries, generating a current that forms the ball. This aerosol model aims to explain the observable properties and behaviors of ball lightning.

Buoyant Plasma Hypothesis: The declassified Project Condign report suggests that ball lightning could be a form of buoyant charged plasma. These plasmas might be created by various atmospheric and electrical phenomena, possibly including meteoroids breaking up in the atmosphere. The report proposes that these plasmas are capable of moving at high speeds, influenced by electrical charges in the atmosphere. However, this explanation has been critiqued as potentially insufficient and not fully encompassing the complexity of the ball lightning phenomenon.

Transcranial Magnetic Stimulation Theory: Cooray and Cooray (2008) noted similarities between the hallucinations experienced by epileptic seizure patients in the occipital lobe and the observed characteristics of ball lightning. They proposed that the rapidly changing magnetic fields from nearby lightning strikes could excite brain neurons, potentially causing hallucinations resembling ball lightning. This theory, however, doesn't account for physical damages associated with ball lightning or its observation by multiple witnesses.

Rydberg Matter Concept: Manykin et al. suggested that ball lightning could be atmospheric Rydberg matter, a condensed form of highly excited atoms. This matter, similar to electron–hole droplets in semiconductors but with a longer lifespan, could form from atmospheric electrical phenomena and appear as a ball lightning event.

Ball lightning, a phenomenon as mesmerizing as it is mysterious, continues to captivate both scientists and the public alike. Despite centuries of sightings and numerous theories ranging from plasma formations to electromagnetic effects, its true nature remains an enigma. The allure of ball lightning lies not just in its striking visual appearance but also in the puzzle it poses to our understanding of atmospheric phenomena. Each hypothesis, from the vaporized silicon theory to the latest in electromagnetic and quantum explanations, adds a piece to this intricate puzzle. As science advances, the mystery of ball lightning remains a fascinating reminder of the wonders and mysteries our natural world still holds, inspiring ongoing research and curiosity.

Earthquake Light

Earthquake lights, alternatively referred to as earthquake lightning or earthquake flashes, present a captivating yet mysterious optical spectacle often observed in the skies above regions experiencing significant tectonic stress, seismic activities, or volcanic eruptions. This phenomenon manifests as an enigmatic illumination, casting an almost ethereal glow in the vicinity of the geological disturbances. Despite numerous studies and observations, the scientific community remains divided, with no broad consensus regarding the underlying causes of this intriguing natural light show. It's important to note that these phenomena are distinct from the disruptions caused to electrical grids. Unlike the bright flashes resulting from arcing power lines, which are typically induced by ground shaking or extreme weather conditions, earthquake lights appear to be a direct byproduct of the Earth's natural geophysical processes, adding another layer of complexity and wonder to our understanding of the planet's dynamic systems.

Earthquake lights have been documented throughout history, with one of the earliest accounts dating back to the 869 Jōgan earthquake in Japan. These lights, often described as unusual illuminations in the sky, have been noted in historical texts like the Nihon Sandai Jitsuroku. Typically, these lights are observed during seismic events, but there have also been instances where they appeared before or after earthquakes, such as the lights reported in relation to the 1975 Kalapana earthquake.

These earthquake lights bear a resemblance to auroras in shape, predominantly displaying a white or bluish hue. However, some reports have described them exhibiting a broader spectrum of colors. The duration of these lights varies, with some witnesses reporting them lasting for several seconds, while others have noted they can persist for up to tens of minutes. The distance from which they can be observed also varies; for instance, during the 1930 Idu earthquake, lights were reported as far as 70 mi from the epicenter.

Notable occurrences of earthquake lights include the 2003 Colima earthquake in Mexico, where colorful lights adorned the skies throughout the earthquake, and the 2007 Peru earthquake, where lights were seen above the sea and captured on film by many. The phenomenon was also documented during the 2009 L'Aquila and the 2010 Chile earthquakes. In New Zealand, during the 1888 North Canterbury earthquake, the lights were visible on the mornings of September 1 and September 8 in Reefton.

In more recent times, earthquake lights have been recorded and widely shared, such as the incidents in Sonoma County, California, on August 24, 2014, and in Wellington, New Zealand, on November 14, 2016, where blue, lightning-like flashes were seen and filmed. On September 8, 2017, many people in Mexico City reported seeing such lights following an 8.2 magnitude earthquake, with the epicenter near Pijijiapan in Chiapas, a significant 460 mi away.

The enigmatic and awe-inspiring phenomenon of earthquake lights seems to favor seismic events of considerable magnitude, typically those registering 5 or higher on the Richter scale. These lights, which manifest in various forms, add a layer of mystical beauty to the otherwise daunting experience of an earthquake. Among these are incidents involving yellow, orb-like illuminations that have been spotted preceding the quakes, adding to the intrigue and mystery surrounding these natural light displays.

Captured on video and shared across the globe, these lights have made a dramatic appearance in recent years. One such instance was during a significant 7.1 magnitude earthquake that shook the city of Acapulco, Mexico, around 20:47 on September 7, 2021. The New York Times reported a stunning visual spectacle as the night sky was lit up with electrical flashes, a dramatic backdrop to the swaying and buckling of power lines in both Acapulco and Mexico City.

Another occurrence was recorded in Qinghai Province, China, at 01:45 on January 8, 2022. A local resident's surveillance video captured a breathtaking moment when the lights made their unexpected appearance. Similarly, during the 2022 Fukushima earthquake in Japan, the phenomenon was documented from multiple vantage points, providing a spectacular visual record of the event.

On September 22, 2022, at around 1:18, a 6.8 magnitude aftershock of the 2022 Michoacán earthquake provided yet another canvas for these mysterious lights. Social media, including Webcams de México, buzzed with videos showing blue lights radiating upwards into the night sky, an arresting sight reported by Mexico News Daily, complete with captivating video evidence.

In a striking display during the 2023 Turkey–Syria earthquake, multiple lights were observed continuously in the provinces of Kahramanmaraş and Hatay, adding an otherworldly aura to the harrowing seismic event. Later that year, the phenomenon made its presence known in Agadir during the Marrakesh–Safi earthquake, where observers were treated to the sight of blue light flashes.

These lights typically manifest themselves a few seconds to several weeks before an earthquake occurs. They are usually observed closer to the epicenter. This type of light display serves as a cryptic, natural warning system, offering a brief and eerie glimpse into the immense tectonic stresses building beneath the Earth's surface.

They can appear either near the epicenter, often as a direct result of earthquake-induced stress, or at considerable distances away from the epicenter, especially during the passage of seismic waves, particularly the S waves. This latter phenomenon is known as "wave-induced stress" and adds a layer of complexity to the understanding of these lights.

Interestingly, earthquake lights appearing during lower magnitude after-shocks are seemingly rare, suggesting a link between the intensity of the seismic event and the likelihood of this phenomenon.

The quest to unravel the mysteries of earthquake lights is an ongoing endeavor in scientific research, with several theories and models proposed to explain their occurrence.

One compelling model suggests that the generation of these lights involves the ionization of oxygen into oxygen anions. This occurs through the breaking of peroxy bonds in certain types of rocks (like dolomite and rhyolite) due to the high stress experienced before and during an earthquake. Post-ionization, these ions ascend through the rock's fissures, reaching the atmosphere. There, they ionize pockets of air to form plasma, which emits the light we see as earthquake lights. Laboratory experiments have supported this theory, demonstrating that some rocks can indeed ionize oxygen under high stress.

Another intriguing hypothesis centers around the intense electric fields generated by the piezoelectric effect. This occurs in quartz-containing rocks, such as granite, due to the tectonic movements. These electric fields could be a key factor in the creation of earthquake lights, adding a fascinating electrical dimension to the geological processes of our planet.

Both these hypotheses, and others in the field, are part of an ongoing scientific investigation, attempting to shed light on these enigmatic and captivating natural phenomena.

One such theory suggests that these lights are the result of local disruptions in the Earth's magnetic field and/or ionosphere in areas of intense tectonic stress. This disruption could potentially result in the observed luminous phenomena, either through radiative recombination in the ionosphere at lower altitudes and higher atmospheric pressure, or as a process akin to auroras. However, it's important to note that this effect is not consistently observed in all earthquake events and has yet to be substantiated through direct experimental verification.

In a fascinating development, the American Physical Society's 2014 March meeting unveiled research that shed light on a possible explanation for these mysterious bright lights observed during earthquakes. The study, led by Troy Shinbrot of Rutgers University, explored the phenomenon by simulating Earth's crust using different types of granular materials. Shinbrot's experiments aimed to replicate the conditions during an earthquake. He discovered that when two layers of the same material rubbed against each other, a voltage was generated. His findings revealed that "when the grains split open, they measured a positive voltage spike, and when the split closed, a negative spike." This process of splitting and closing allowed the voltage to discharge into the air, electrifying it and consequently producing a bright electrical light.

This revelation was consistent across all materials tested, suggesting a universal principle at play. Shinbrot drew parallels to the phenomenon of triboluminescence, a process where light is emitted through the breaking of chemical bonds in a material when it is pulled apart, ripped, scratched, crushed, or rubbed. The implications of this research extend far beyond the academic curiosity surrounding earthquake lights. By delving deeper into this phenomenon, scientists hope to glean crucial insights that could significantly enhance our understanding of seismic activities. This, in turn, could pave the way for advancements in earthquake prediction, potentially saving lives and mitigating the impact of these natural disasters. Shinbrot's research represents a pivotal step in unraveling the mysteries of the Earth's geophysical processes, bringing us closer to comprehending and possibly foreseeing the formidable forces of nature.

St. Elmo's Fire

St. Elmo's fire, a captivating phenomenon, stands as a testament to nature's enigmatic beauty. This mesmerizing form of plasma, not only reproducible but also demonstrable, emerges under specific atmospheric conditions, primarily during thunderstorms. It's the result of an electric field around an object ionizing the surrounding air molecules. This ionization process emits a faint yet discernible glow, particularly visible in low-light conditions, akin to a mystical aura.

The genesis of St. Elmo's fire requires a local electric field intensity of approximately 100 kV/m, typically arising in moist air during thunderstorms where high-voltage differentials exist between the clouds and the earth. The formation and intensity of this electric field are heavily influenced by the geometry of the object involved, including its shape and size. Interestingly, objects with sharp points necessitate a lower voltage to initiate a discharge. This is due to electric fields being more densely concentrated in areas with high curvature. Consequently, discharges are more prone to occur and appear more intense at the tips of pointed objects, creating a striking visual effect.

The unique blue or violet glow of St. Elmo's fire results from the nitrogen and oxygen in Earth's atmosphere. This phenomenon is akin to the mechanism behind the glow of neon lights, although the colors differ due to the distinct gases involved. It's a captivating visual display of nature's interaction with atmospheric elements.

Historically, this phenomenon has intrigued scientists and observers alike. In 1751, the renowned Benjamin Franklin hypothesized that during a lightning storm, a pointed iron rod would exhibit a glow at its tip, similar in appearance to St. Elmo's fire. This early observation marked a pivotal point in our understanding of atmospheric electricity.

More recently, in August 2020, a groundbreaking paper by researchers from MIT's Department of Aeronautics and Astronautics shed new light on the behavior of St. Elmo's fire. Their research revealed distinct differences in how St. Elmo's fire manifests on airborne objects compared to grounded structures. They discovered that electrically isolated structures, especially in high winds, accumulate charge more effectively, contrasting with the corona discharge observed in grounded structures. This revelation opened new avenues in understanding atmospheric electrical phenomena and their impact on various structures, both in the air and on the ground.

In ancient Greece, a single instance of this ethereal glow was known as Helene, translating to "torch" in Ancient Greek, symbolizing a solitary beacon in the dark. When it appeared as a pair, it was named Castor and Pollux, after the mythological twin brothers of Helen, imbuing it with a sense of celestial mystique.

Post-medieval interpretations of St. Elmo's fire saw it being linked with elemental forces. Influential figures like Paracelsus associated it with elementals such as the salamander, a mythical creature of fire, or an acthnici, a similarly enigmatic entity. This association with mythical creatures rooted it deeply in the lore of elemental magic and alchemy.

Maritime cultures had their own interpretations and names for St. Elmo's fire. Welsh mariners called it canwyll yr ysbryd or canwyll yr ysbryd glân, translating to "candles of the Holy Ghost" or the "candles of St. David," reflecting a spiritual reverence. Russian sailors, documenting instances of this phenomenon, referred to it as "Saint Nicholas" or "Saint Peter's lights," and sometimes as St. Helen's or St. Hermes' fire, possibly due to linguistic mix-ups. These names highlighted the phenomenon's significance in seafaring lore as symbols of protection or omens.

The historical significance of St. Elmo's fire is marked by its appearance during crucial moments. One such instance was during the Siege of Constantinople in 1453, where it was seen emitting from the top of the Hippodrome. The Byzantines saw this as a divine sign of impending victory against the Ottoman Empire, a belief that was shattered when Constantinople fell shortly after the light's disappearance.

St. Elmo's fire also graced the voyages of explorers like Magellan. During his first circumnavigation of the globe, the phenomenon, referred to as the body of St. Anselm, was seen multiple times around the fleet's ships off the South American coast, interpreted by sailors as favorable omens.

In more recent history, St. Elmo's fire continued to be observed in remarkable circumstances. Notably, the B-29 Bockscar, carrying the Fat Man atom bomb to Nagasaki in 1945, experienced a mysterious blue plasma around its propellers, described as a "chariot of blue fire." Similarly, during the 1955 Great Plains tornado outbreak, this natural light show made an appearance, adding a spectral dimension to this dramatic event.

In aviation, St. Elmo's fire has been both observed and mistaken for other phenomena. For example, on British Airways Flight 9 in 1982, glowing light flashes along the aircraft's leading edges were initially thought to be St. Elmo's fire but were later identified as the impact of ash particles, akin to sandblasting effects. Furthermore, during a University of Alaska research flight over the Amazon in 1995, St. Elmo's fire was not only observed but its optical spectrum was recorded, contributing to the study of sprites.

Even in tragic events, such as the crash of Air France Flight 447 in 2009, St. Elmo's fire played a role, being observed 23 minutes before the crash, although not contributing to the disaster.

In Filipino folklore, St. Elmo's fire, known as Apoy ni San Elmo or santelmo, is seen as a bad omen or a flying spirit. This interpretation, more akin to ball lightning, predates the term santelmo, which originated during Spanish colonial rule in the Philippines. The term reflects a blend of indigenous beliefs and colonial influences, demonstrating the phenomenon's deep cultural resonance across the globe.

Julius Caesar in "De Bello Africo" and Pliny the Elder in "Naturalis Historia" both reference this enigmatic glow. Alcaeus, an ancient Greek lyric poet, also alludes to it, as does Xenophanes of Colophon, a philosopher and poet, indicating the phenomenon's deep roots in ancient thought and literature.

In the 15th century, the famed Ming Dynasty Admiral Zheng He recorded a fascinating encounter with St. Elmo's fire during his treasure voyages. In the Liujiagang and Changle inscriptions, he and his associates described witnessing this celestial light as a divine omen from Tianfei, the goddess revered by sailors and seafarers. They recounted how the sudden appearance of a shining light at the masthead during a hurricane calmed the raging storm, providing reassurance and safety. This account, part of the Changle inscription, highlights the spiritual and mystical importance attributed to St. Elmo's fire in different cultures.

The phenomenon also finds mention in the explorations of Ferdinand Magellan and Vasco da Gama. Antonio Pigafetta, a chronicler of Magellan's voyage, documented instances of St. Elmo's fire, known in Portuguese as "corposants" or "corpusants," meaning "holy body." Similarly, "The Lusiads," the epic poem detailing Vasco da Gama's voyages, describes this mysterious light, reinforcing its significance in the annals of maritime exploration.

In the early 17th century, Robert Burton, in his "Anatomy of Melancholy," recounts the experience of Mikołaj Krzysztof "the Orphan" Radziwiłł, a

Lithuanian duke. Radziwiłł, during a voyage from Alexandria to Rhodes in 1582, observed St. Elmo's fire, referring to it as "Sancti Germani sidus," indicating its perceived celestial nature.

John Davis, the renowned English navigator, also encountered St. Elmo's fire during his second voyage to the East Indies in 1605. Aboard the Tiger, an unknown writer described a flame appearing on the main topmast head during a severe storm. This sighting, referred to as Corpo Sancto by the Portuguese, was seen as a divine sign indicating the passing of the worst part of the storm, a belief that was seemingly validated by the subsequent improvement in weather.

Pierre Testu-Brissy, a French balloonist known for his pioneering flights, had a significant encounter with St. Elmo's fire on 18 June 1786. During an 11-hour flight, he ascended into thunderclouds and made groundbreaking electrical observations. Using an iron rod carried in the balloon's basket, he noted remarkable electrical discharges from the clouds and experienced the mystical glow of St. Elmo's fire, marking a noteworthy moment in the intersection of exploration and scientific observation.

William Bligh, the infamous captain of HMS Bounty, also recorded an encounter with St. Elmo's fire. On 4 May 1788, he noted in his log the sight of electrical vapor around the ship's iron yardarms, resembling the blaze of a candle. This observation was made in the South Atlantic, as the Bounty journeyed from Cape Horn towards the Cape of Good Hope, highlighting the phenomenon's prevalence in maritime explorations.

William Noah, a silversmith turned convict, provided a detailed account of St. Elmo's fire in his journal during his transport to Sydney, New South Wales, on the Hillsborough. On 26 June 1799, amid a heavy storm in the Southern Ocean, he described a corposant, a ball of fire, which appeared on the foretopmast head and subsequently burst on the main deck. Another sighting on 25 July 1799 in the Tasman Sea led to two seamen being temporarily blinded. These dramatic incidents illustrate the awe and fear that St. Elmo's fire could inspire during sea voyages.

In 1817, James Braid, a surgeon, experienced St. Elmo's fire during an electrical storm in Lanarkshire. While on horseback, he observed the tips of the horse's ears and the edges of his hat becoming luminous. This fleeting and beautiful occurrence was accompanied by numerous minute sparks and highlighted the atmospheric conditions conducive to St. Elmo's fire.

Earlier, in January 1817, a unique weather event in Vermont and New Hampshire featured St. Elmo's fire during a luminous snowstorm. Static discharges were observed on roof peaks, fence posts, and even on people's hats and fingers, accompanied by prevalent thunderstorms over central New England. This event stands as a testament to the widespread and varied manifestations of St. Elmo's fire in different environmental conditions.

One such notable observer was Charles Darwin, who experienced this phenomenon while aboard the HMS Beagle. In a letter to J. S. Henslow, Darwin described a night in the estuary of the Río de la Plata where the sky and water seemed ablaze with lightning and luminous particles, and even the ship's masts were tipped with a blue flame. He elaborated on this in his book "The Voyage of the Beagle," recounting a night of natural fireworks where St. Elmo's light shone from the mast-head and yard-arm-ends, creating a scene so vivid that the sea, illuminated by luminous particles, highlighted the paths of penguins with a fiery wake.

Richard Henry Dana Jr., in his book "Two Years Before the Mast," describes witnessing a corposant in the horse latitudes of the northern Atlantic Ocean. He observed a ball of light, known to sailors as a corposant or corpus sancti, on the main top-gallant mast-head. Sailors believed that the movement of this light in the rigging could predict the weather, viewing it as an omen of either fair weather or an impending storm.

Nikola Tesla, a pioneer in electrical engineering, successfully created St. Elmo's fire in his experiments. In 1899, while testing a Tesla coil at his laboratory in Colorado Springs, he observed St. Elmo's fire around the coil, which even illuminated the wings of butterflies with blue halos as they flew around.

Mark Heald, a professor at Princeton, witnessed St. Elmo's Fire just before the tragic crash of the LZ 129 Hindenburg in 1937. He saw a dim "blue flame" flickering along the airship's backbone, a sight that preceded the catastrophic burst of flaming hydrogen that led to the airship's demise.

William L. Laurence, a reporter for The New York Times, reported seeing St. Elmo's fire on 9 August 1945, while aboard the Bockscar en route to Nagasaki. He described a strange light coming through the window and saw luminous discs of blue flame around the propellers and on the aircraft's windows and wingtips. Captain Bock, who was piloting the Bockscar, reassured him, explaining that it was a familiar phenomenon known as St. Elmo's Fire, often seen on ships and aircraft.

Will-o'-the-wisp

In the realm of folklore, the enigmatic will-o'-the-wisp, also known as will-o'-wisp or ignis fatuus—a Latin term translating to 'foolish flame'—presents a captivating yet eerie spectacle. This atmospheric ghost light, typically observed by travelers during the dark hours of the night, is most commonly seen hovering over bogs, swamps, or marshes. This phenomenon, deeply embedded in European folklore, goes by numerous names across different cultures, including the jack-o'-lantern, friar's lantern, and hinkypunk. It is famously reputed for leading unwary travelers astray, appearing as a deceptive, flickering lamp or lantern, luring them into uncharted territories.

Beyond its literal interpretation, in literature, the will-o'-the-wisp has taken on a metaphorical meaning. It often symbolizes an elusive hope or goal, tantalizingly out of reach, or it might represent something that is both intriguing and somewhat ominous or sinister. These mystical lights have made their way into folk tales and traditional legends of various countries and cultures, with each having its notable examples. Among these are the St. Louis Light in Saskatchewan, the Spooklight in Southwestern Missouri and Northeastern Oklahoma, the enigmatic Marfa lights of Texas, the Naga fireballs along the Mekong in Thailand, the Paulding Light in Michigan's Upper Peninsula, and the Hessdalen light in Norway.

In folklore, wills-o'-the-wisp are often attributed to otherworldly beings like ghosts, fairies, or elemental spirits, adding a layer of supernatural intrigue to their lore. However, modern scientific explanations tend to demystify these lights, attributing their occurrence to natural phenomena such as bioluminescence or chemiluminescence. This is believed to be caused by the oxidation of certain gases like phosphine, diphosphane, and methane, which are byproducts of organic decay. This scientific explanation provides a fascinating contrast to the rich tapestry of myths and legends that have surrounded these mysterious lights for centuries.

The captivating folklore surrounding "hobby lanterns," a term originating from the 19th century Denham Tracts, reveals a rich tapestry of beliefs and tales linked to this mysterious phenomenon. In her seminal work, A Dictionary of Fairies, K. M. Briggs delves deeply into this subject, offering a comprehensive catalogue of various names ascribed to these enigmatic lights, which vary significantly based on their geographical locations, such as graveyards or bogs. In graveyards, they are eerily referred to as "ghost candles," another intriguing nomenclature drawn from the Denham Tracts.

These ethereal lights also feature prominently in folk tales across various cultures, including Ireland, Scotland, England, Wales, Appalachia, and Newfoundland. These narratives often involve protagonists, typically named Will or Jack, who are cursed to perpetually wander marshlands, bearing a lantern as punishment for their misdeeds. A particularly poignant version from Shropshire, recounted in Briggs' A Dictionary of Fairies, tells of Will Smith, a nefarious blacksmith. Despite being granted a chance for redemption by Saint Peter at heaven's gates, Will's continued wickedness condemns him to an eternal earthbound wandering. The Devil, taking pity, offers him a solitary burning coal for warmth, which Will then maliciously employs to lure unwary travelers into treacherous marshes.

In an Irish variation of the tale, the central figure is a mischievous character known as Drunk Jack or Stingy Jack. Faced with the Devil coming to claim his soul, Jack cunningly tricks the Devil into transforming into a coin to settle his last drink. Once the Devil complies, Jack traps him in his pocket beside a crucifix, inhibiting his return to his original form. The Devil, in exchange for his freedom, extends Jack's life by ten years. Upon the expiration of this term, Jack once again deceives the Devil, this time by forcing him up a tree and carving a cross underneath, thus trapping him. In return for removing the cross, the Devil absolves Jack of his debt. However, given Jack's nefarious deeds, heaven denies him entry upon his death. When Jack turns to hell for refuge, the Devil, in a final act of revenge, refuses him entry but grants him an ember from hell's fires to light his path in the eternal twilight world to which lost souls are doomed. Jack uses this ember in a hollowed-out turnip, creating his own lantern. This story is just one of many variants, including "Willy the Whisp" in Irish Folktales by Henry Glassie and Séadna by Peadar Ua Laoghaire, notable for being the first modern novel written in the Irish language.

In Mexico, these ghostly lights are enveloped in legends, often described as witches who have undergone a mysterious transformation. Another captivating aspect of Mexican folklore suggests that these lights, known as "luces del dinero" (money lights) or "luces del tesoro" (treasure lights), are mystical indicators, pointing towards buried treasures or gold, which, according to legend, can only be unearthed with the innocence and purity of children.

Venturing to the swampy regions of Massachusetts, particularly the Bridgewater Triangle, one encounters tales of spectral orbs of light. These ghost-lights, having been observed even in modern times, add a layer of contemporary mystique to the ancient folklore, suggesting a bridge between the past and present in these mysterious occurrences.

In Louisiana, the legend of the fifollet (or feu-follet) emerges from French cultural influences. This tale speaks of a tormented soul, returned from the dead to perform penance as decreed by God. However, this soul, consumed by vengeance, often engages in mischievous or malevolent acts, including, in some darker versions of the tale, the bloodsucking of children. It is said that the fifollet could be the unsettled spirit of an unbaptized child.

Brazil's folklore introduces Boi-tatá, a name derived from the Old Tupi language, translating to "fiery serpent" (mboî tatá). This mythical creature, not quite a dragon but more akin to a colossal snake, is said to have survived a great deluge. Emerging from its cave, the Boi-tatá preyed upon animals and corpses, devouring their eyes, which in turn imbued it with its fiery gaze. By day, its vision is impaired by the brightness it has accumulated, but by night, it sees all.

The lore of Argentina and Uruguay brings forth the concept of "luz mala" (evil light), a phenomenon deeply ingrained in the folklore of these lands. Often witnessed in rural areas, it appears as a dazzling ball of light, hovering eerily above the ground, instilling fear and awe among those who encounter it.

In Colombia, the tale of la Bolefuego or Candileja presents a ghostly will-o'-the-wisp that is the spirit of a morally corrupt grandmother, whose life led her grandchildren down the path of thievery and murder. In her afterlife, she is condemned to an eternal, fiery wandering. Similarly, in Trinidad and Tobago, the legend of the soucouyant depicts a "fireball witch," a malevolent spirit transforming into a flame at night, slipping through the smallest of gaps to enter homes and feast on the blood of its victims.

The Aleya lights, also known as marsh ghost-lights, present a mesmerizing and eerie phenomenon observed over the marshes of Bangladesh and West Bengal. These mysterious lights, often seen by local fishermen, are said to be manifestations of marsh gas apparitions. They carry a dual nature in local lore: while they sometimes lead fishermen astray, causing them to lose their bearings and potentially meet a watery grave, they are also believed to be protective spirits. These ghost-lights are thought to represent the souls of deceased fishermen, guiding their living counterparts away from impending dangers or confusing them with their deceptive dance over the marshes.

In the diverse and mystical landscapes of India, similar ghost-lights captivate the imagination. Known as Chir batti in the Banni grasslands and the adjoining Rann of Kutch's desert, these dancing lights illuminate the dark nights, creating an almost otherworldly spectacle. Other regional variants like the Kollivay Pey of Tamil Nadu and Karnataka, the Kuliyande Choote of Kerala, and numerous forms from the tribes of Northeast India add rich layers to India's folkloric tapestry surrounding these enigmatic lights.

Japanese folklore, too, is replete with tales of similar phenomena. Hitodama, translating to "Human Soul," and Hi no Tama, or "Ball of Flame," are often associated with graveyards, hinting at a connection with the afterlife. The Kitsune, mythical yokai demons, are intertwined with these lights; the union of two Kitsune is said to produce kitsune-bi, or "fox-fire," a fascinating blend of mythology and natural phenomenon. These stories are vividly depicted in Shigeru Mizuki's book "Graphic World of Japanese Phantoms."

In Korea, these mysterious lights, known as 'dokkebi bul' (goblin fire), are often seen around rice paddies, ancient trees, and mountains, and sometimes even in houses. Considered both malevolent and mischievous, they are believed to lead unsuspecting travelers astray, causing them to lose their way or fall into hidden dangers in the dead of night.

The Chinese historical texts provide some of the earliest references to the will-o'-the-wisp. The character lín, found in the Shang dynasty oracle bones, depicts a human-like figure surrounded by dots, symbolizing the glowing lights. Over time, this character evolved, incorporating elements that represented fire and later rice. The element phosphorus, scientifically linked to the will-o'-the-wisp phenomenon, is found in the character lín, a testament to the deep historical roots of these tales in Chinese culture. Shen Gua, a Chinese polymath, recorded a notable occurrence in his Book of Dreams, describing an enormous, bright pearl appearing in the marshes of Yanzhou, Jiangsu province. This light was so reliable that it persisted for over a decade, leading locals to construct the Pearl Pavilion for observers to marvel at this natural wonder.

In Sweden, the will-o'-the-wisp takes on a particularly heartrending character. It is thought to represent the soul of an unbaptized individual, eternally seeking baptism. According to legend, these souls attempt to lead travelers towards water, hoping against hope for the baptism they never received in life, a ritual believed to grant them peace and passage to the afterlife.

Across Denmark, Finland, Sweden, Estonia, Latvia, Lithuania, Ireland, and other cultures, the will-o'-the-wisp is often intertwined with tales of hidden treasures. These ghostly lights are said to mark the spot where vast riches lie buried deep within the earth or beneath water, awaiting discovery. However, unearthing these treasures is no simple task. Legends speak of the need for magical rituals, and sometimes even the use of a dead man's hand, to access these hidden riches. In Finland and other northern lands, folklore dictates that early autumn, particularly around the time of the summer solstice or Midsummer's Day, is the prime time to seek out these elusive lights and the treasures they guard. The lore suggests that those who buried these treasures long ago did so with enchantments, ensuring that they could only be reclaimed under the ethereal glow of the will-o'-the-wisp on specific, enchanted nights.

Delving deeper into Finnish mythology, we encounter the Aarnivalkea, also known as virvatuli, aarretuli, and aarreliekki. These are not mere flickers of light but spots where an eternal flame burns, marking where the fabled faerie gold is buried. Shrouded in enchantments, these treasure spots are hidden from the eyes of the casual seeker. Only those fortunate enough to find a fern seed from the mythical flowering fern can hope to discover these treasures. This fern, which in reality does not bloom and reproduces through spores, is said in myth to bloom under the rarest of circumstances. The finder of this seed is not only guided to the hidden riches but is also bestowed with a cloak of invisibility, enabling them to move unseen through the world.

Welsh folklore is rich with tales of the mysterious and the magical, among which is the legend of "fairy fire," a phenomenon attributed to the mischievous deeds of a púca or pwca. This goblin-like fairy, part of the Tylwyth Teg or fairy family, is known for leading unwary travelers astray with a flickering light held in its hand. As these solitary wanderers follow the púca through marshes or bogs, the light is suddenly extinguished, leaving them disoriented and lost in the darkness.

A captivating tale mentioned by Wirt Sikes in his book "British Goblins" illustrates this trickery. A peasant, journeying home at dusk, spots a bright light moving ahead of him. Upon closer inspection, he realizes it's a lantern carried by a shadowy, diminutive figure. He follows this figure for miles, only to find himself on the brink of a vast chasm with a torrential river below. In a sudden twist, the lantern-carrier leaps over the gap, raises the lantern high, emits a spiteful laugh, and extinguishes the light, abandoning the peasant in complete darkness on the precipice.

In some Welsh beliefs, this light also harbors an ominous prediction, foretelling an imminent funeral in the local area. However, the will-o'-the-wisp isn't always seen as malevolent. Some stories suggest that these lights are guardians of treasure, akin to the Irish leprechaun, leading the daring to fortunes. Other narratives tell of lost travelers encountering a will-o'-the-wisp in the woods, with their fate - further misdirection or guidance to safety - depending on how they interact with the spirit.

Related to these tales are the "pixy-lights" from Devon and Cornwall, known for misleading travelers from safe paths into treacherous bogs. These lights are akin to mischievous poltergeists, often playing pranks like blowing out candles of courting couples or creating misleading sounds. In Cornish folklore, the pixy-light is associated with the "Colt pixie," a shape-shifting pixie that assumes the form of a horse, leading other horses astray with its neighing.

In Guernsey, the phenomenon is known as the "faeu boulanger" (rolling fire), believed to be a wandering lost soul. Folk remedies to confront this apparition include turning one's cap or coat inside out to halt its progress or placing a knife blade-up in the ground, which the faeu would then attack in an attempt to destroy itself.

In the Scottish Highlands, the will-o'-the-wisp is known as the "Spunkie," often appearing as a linkboy or a receding light, leading travelers to peril. It has even been blamed for causing shipwrecks at night, being mistaken for a harbour light. Scottish folklore also regards these lights as omens of death or the spirits of the deceased, appearing over lochs or along roads frequented by funeral processions. Additionally, a peculiar light seen in the Hebrides, known as the "teine sith" or "fairy light," though not directly connected to the fairy race, adds another layer to the rich tapestry of Celtic and Scottish myths surrounding these enigmatic lights.

The will-o'-the-wisp phenomenon, also known as ignis fatuus, has long been a subject of both folklore and scientific curiosity. Modern science explains these mysterious lights as a result of chemical reactions involving natural gases. Specifically, the oxidation of phosphine, diphosphane, and methane, which are by-products of organic decay, can lead to the emission of photons, thus creating these ephemeral fires. Notably, phosphine and diphosphane can spontaneously ignite upon contact with oxygen, providing a potential ignition source for the more abundant methane to produce these fleeting flames.

Additionally, the production of phosphorus pentoxide as a byproduct of phosphine oxidation, which forms phosphoric acid when it contacts water vapor, could account for the "viscous moisture" occasionally reported in association with ignis fatuus sightings.

The concept of will-o'-the-wisp being linked to natural gases dates back to historical texts. Ludwig Lavater, in 1596, is among the earliest to mention this idea. In 1776, Alessandro Volta proposed that ignis fatuus could result from natural electrical phenomena, like lightning, interacting with marsh gas, notably methane. This hypothesis was later supported by Joseph Priestley, a

British polymath, in his series of works titled "Experiments and Observations on Different Kinds of Air" (1772–1790), and by French physicist Pierre Bertholon de Saint-Lazare in his work "De l'électricité des météores" (1787).

However, early critics of the marsh gas hypothesis raised several objections, including doubts about spontaneous combustion, the absence of warmth in some ignis fatuus sightings, and the peculiar behavior of these lights receding when approached. Critics also differentiated ignis fatuus from ball lightning, another unexplained natural phenomenon. An example of such criticism is found in John G. Owens' "Folk-Lore from Buffalo Valley" (1891).

The tendency of ignis fatuus to retreat when approached might be simply explained by the disturbance of air currents. When air is agitated by moving objects, it could cause the gas to disperse. This theory was substantiated by the detailed observations of Major Louis Blesson, who conducted experiments in the 1830s in various known locations of ignis fatuus sightings. He particularly noted a marshland in Gorbitz, Newmark, Germany, where he observed bluish-purple flames that receded whenever he attempted to approach them. Through persistent efforts, Blesson confirmed that these lights were indeed caused by ignited gas. British scientist Charles Tomlinson later described Blesson's experiments in "On Certain Low-Lying Meteors" (1893).

Blesson also noted variations in the color and heat of the flames in different marshes. For instance, ignis fatuus observed in Malapane, Upper Silesia (now Ozimek, Poland) could be ignited and extinguished without burning paper or wood shavings. In another Polish forest, the ignis fatuus coated objects with an oily viscous fluid instead of burning them. Interestingly, Blesson accidentally created ignis fatuus while setting off fireworks in the marshes of Porta Westfalica, Germany, further supporting the theory of their chemical origin.

In 1980, British geologist Alan A. Mills from Leicester University made a notable attempt to scientifically recreate the ignis fatuus phenomenon in a laboratory setting. While Mills did manage to produce a cool glowing cloud by mixing crude phosphine and natural gas, the resulting light was green and accompanied by a significant amount of acrid smoke. This outcome contrasted starkly with most eyewitness descriptions of ignis fatuus, which typically don't mention such smoke or a greenish hue. In response to these discrepancies, Mills proposed in 2000 that the natural occurrence of ignis fatuus might instead be explained by cold flames, a type of luminescent pre-combustion halo that occurs when certain compounds are heated just below their ignition point. Notably, cold flames, often bluish, emit very little heat and can occur in a range of substances, including many that are natural byproducts of organic decay.

Building on Mills' research, Italian chemists Luigi Garlaschelli and Paolo Boschetti in 2008 attempted to replicate these experiments. They succeeded in creating a faint, cool light by mixing phosphine with air and nitrogen. Although the light they produced was still greenish, they pointed out that under low-light conditions, human eyes might not easily distinguish colors. By adjusting gas concentrations and environmental factors like temperature and humidity, they could reduce or eliminate the smoke and smell. Garlaschelli and Boschetti concurred with Mills that cold flames could be a plausible explanation for some instances of ignis fatuus.

Another interesting theory was put forward in 1993 by professors Derr and Persinger, who suggested that some ignis fatuus phenomena might have a geological origin, being piezoelectrically generated under tectonic strain. This hypothesis posits that the strains which move faults could heat rocks, vaporizing water within them. If these rocks or soils contain piezoelectric materials like quartz, silicon, or arsenic, they could produce electricity. This electricity might then travel up to the surface through a column of vaporized water, manifesting as mysterious earth lights, and possibly explaining the seemingly electrical and erratic behavior of ignis fatuus.

Additionally, bioluminescence in forest microorganisms and insects might also account for some will-o'-the-wisp sightings. The natural glow from certain fungi, such as the honey fungus, during their chemical processes could be mistaken for ignis fatuus or foxfire lights. Furthermore, light reflections from larger forest creatures or bioluminescent organisms like fireflies could create the illusion of moving fairy lights. The reflective white plumage of barn owls, for instance, might appear as moving lights under moonlight, adding to the lore of these phenomena.

The decline in ignis fatuus sightings in recent times is often attributed to the draining and reclamation of swamps and marshlands, such as the once extensive Fenlands of eastern England, now transformed into farmland. This environmental change has likely reduced the conditions favorable for the natural occurrence of these mysterious lights.

Bibliography

Barry, James Dale. "Ball Lightning and Bead Lightning." Plenum Press, 1980.

Behe, George, and Michael Goss. "Lost at Sea: Ghost Ships and Other Mysteries." Prometheus Books, 2005.

Brown, Alan. "Haunted Places in the American South." University Press of Mississippi, 2002.

Darack, Ed. "Unlocking the Atmospheric Secrets of the Marfa Mystery Lights." Weatherwise, 2008.

Goodrich, Marcia. "Just in Time for Halloween: Michigan Tech Students Solve the Mystery of the Paulding Light." Michigan Tech, 28 Oct. 2010, https://www.mtu.edu/news/2010/10/just-time-for-halloween-michigan -tech-students-solve-mystery-paulding-light.html.

Joiner, G. D. "Historic Haunts of Shreveport." History Press, 2010.

Kozicka, Maureen. "The Mystery of the Min Min Light." M Kozicka, 1994.

MacGregor, Alasdair Alpin. "The Peat Fire Flame: Folktales and Traditions of the Highlands and Islands." Moray Press, 1937.

Mitford, A.B. "Japanese Legends and Folklore: Samurai Tales, Ghost Stories, Legends, Fairy Tales, Myths and Historical Accounts." Tuttle Publishing, 2019.

Randolph, Vince. "Ozark Magic and Folklore." Dover Publications, 1964.

Ross, Catrien. "Haunted Japan: Exploring the World of Japanese Yokai, Ghosts and the Paranormal." Tuttle Publishing, 2020.

Sherwood, Roland H. "The Phantom Ship of Northumberland Strait And Other Mysteries Of The Sea." Lancelot Press, 1975.

Stewart, William Grant. "The Popular Superstitions and Festive Amusements of the Highlanders of Scotland." 1823.

Stollznow, Karen. "Haunting America: The Truth Behind Some of America's

'Most Haunted' Places." Amazon Books, 2020.

Strömberg, Sulo. "Kerimäen ja Savonrannan kyliä kiertämässä. Tarinoita Paasveen piruista ja pohuista." 2006.

Theakston, Oliver. "St. Elmo's Fire: A Cursed Voyage of Discovery That Changed the World and Destroyed the Men Who Undertook It." Independent, 2022.